CLEAN AIR
ACTIVITIES

CLEAN AIR ACTIVITIES

Exercises for a Cleaner Tomorrow

by The Clean Air Conservancy

Humanics Learning
Atlanta, GA USA

HUMANICS LEARNING

Clean Air Activities
A Humanics Learning Publication

Humanics Learning Publications are an imprint of and published by Humanics Publishing Group, a division of Brumby Holdings, LLC. It's trademark, consisting of the words "Humanics Learning" and a portrayal of a silhouetted girl, is registered in the U.S. Patent Office and in other countries.

Brumby Holdings, LLC
1197 Peachtree Street
Suite 533 Plaza
Atlanta, GA 30361

Printed in the United States of America

Library of Congress Control Number: 2002110154
ISBN (Paperback): 0-89334-377-3
ISBN (Hardcover): 0-89334-378-1

table of contents

3

Introduction

7

Getting Started

23

Topic Outlines

31

Teacher Resources

75

Experiments

88

Student Information

99

Student Workbook

1

a c k n o w l e d g e m e n t s

This curriculum is the outcome of over four years of interaction between students, teachers, and the Conservancy staff. Given that there are many people who have made large contributions to the development of this package, our thanks far exceeds our ability to acknowledge them. For the record, this is at best a partial list of all those who deserve our thanks.

Rod Johnson
Norm Schmidt
Charlotte Freeman
Pat Carpenter
Robyn Cella Francis
Peter Thomas
Stand and Hope Adelstein
The Board of the Clean Air Conservancy
The George Gund Foundation
The William Bingham Foundation
The National Parks and Conservation Association
The Ohio Environmental Education Foundation
The Brandon Family Foundation
The Raymond Wean Foundationing

I would also like to offer a special thanks to Jayne Cella Miller and Jessica Dunn, the education and outreach arms of the Conservancy, for the excellence of their work. I would also like to thank Cynthia Van Lenten, our graphic designer, who was instrumental in makthis package a visual success.

While all of those mentioned, and many others, deserve our thanks, this package is ultimately the responsibility of the Clean Air Conservancy. Any errors, omissions, or mis-interpretation of data is solely the responsibility of the Conservancy staff.

Kevin F. Snape, Ph.D.
Executive Director
The Clean airConservancy

Purpose

The main purpose of this curriculum module is to introduce students to the causes of air pollution and its impacts. There is a focus on the air pollution generated by coal burning electrical utilities, however, there is some attention on other sources of air pollution. The module also offers hands-on solutions for lessening pollution. Students will explore the causes of air pollution and be made aware of the amount of pollution in the air. Students will also learn ways in which they can help the environment.

The pollution emitted by power companies is full of sulfur dioxide (SO_2), nitrogen oxide (NO_x), and carbon dioxide (CO_2), as well as other pollutants. These gases and particles are in our air and ultimately end up in our water. By gaining an understanding of the connection between electricity and pollution, students will have a greater awareness of their role in creating smog and acid rain. This module is designed to provide information to students so that they might find methods for limiting and controlling air pollution.

The Clean Air Conservancy has designed this module to be used as a series of topics. Many of these topics have elements that are relevant to other areas of study. Earth science, biology, chemistry, mathematics, and economics are some of the other lessons taught through this module. While it was designed to be taught as a series, the individual topics can be done separately. The user can take pieces of this module and incorporate them into their own lesson plans.

However you choose to use this information, it is the hope of the Clean Air Conservancy that, through education and exploration, students will come to understand the importance of these topics and their relevance to everyday life. If students are able to make the connection between the generation of electricity and the emission of pollutants they will then be ready to take the necessary steps to make a difference.

the clean air conservancy

Outline

This module is designed around five major topics:

1. The Different Pollutants and Their Sources
2. Environmental Effects of Air Pollution
3. Government Regulation: A Way to Control Air Pollution
4. Alternative Ways to Control Air Pollution: Energy Efficiency, Alternative Electricity Generation
5. Using Technology to Control Pollution

Each topic has an icon so that you can easily connect different elements that compliment one another.

1. Getting Started

This section contains material for the instructor. We have provided this section as research for your teaching. The information is intended as a starting point in your research. Further, the Conservancy web page (www.cleanairconservancy.org) provides more information on each topic as well as an opportunity to e-mail questions to us. The web site also offers other websites that can be used to further research the different topics. Although designed as information for instructors, you might find that some of the information is useful to pass on to your student.

2. Topic Outlines

Provides procedure, focus, methods and conclusions for each topic. This section is intended to help you create lesson plans around the different topics.

3. Teacher Resources

This section provides much of the needed material for doing some of the different activities suggested in the Topic Outlines. The answer keys for the worksheets are also in this section.

4. Experiments

All of the different experiments suggested in the Topic Outlines are found in this section. You can easily connect the proper experiments and topics using the icons. There are also other experiments you might want to use.

5. Student Information

Some of the activities have information sheets that will provide students with valuable background information. There are also some other ideas and reference materials that are provided in this section.

6. Student Workbook

This section has a worksheet and some glossary terms for each different topic. It is designed so that you can assess students' comprehension of these topics.

How to Use

On the top of each page is an icon which represents the major theme of that page. Some pages have more than one icon since they can be used for more than one topic. This is done so that you might quickly connect all the different pieces which are relevant to one topic.

1. The Different Pollutants and Their Sources

The factory and smokestacks icon represents the pollution that factories create during production. The student worksheets, topic outline, background reading and materials relevant to this topic all include this icon.

2. Environmental Effects of Air Pollution

The mountain and clouds icon represents the environment. The student worksheets, topic outline, background reading and materials relevant to this topic all include this icon.

3. Government Regulation: A Way to Control Air Pollution

The eye icon represents those who are watching the polluters to make sure that they are doing what they have agreed to do. The student worksheets, topic outline, background reading and materials relevant to this topic all include this icon.

4. Alternative Ways to Control Air Pollution: Energy Efficiency, Alternative Electricity Generation

The windmill icon represent other ways, besides burning coal that electricity can be generated. The student worksheets, topic outline, background reading and materials relevant to this topic all include this icon.

5. Using Technology to Control Pollution

The broom icon represents cleaning up the pollution before it enters the air. The student worksheets, topic outline, background reading and materials relevant to this topic all include this icon.

the clean air conservancy

The bird flying through clean air represents material which can be used as a conclusion or tie together many of the different topics.

The icons are intended to make using this unit very easy. Hopefully, by matching up the different icons, you can create lesson plans and teaching units that are useful to you and your students.

Air Pollution

SULFUR DIOXIDE (SO$_2$)

Sulfur dioxide is largely a byproduct of burning fossil fuels, especially coal. The sulfur is trapped in the coal as it forms (along with a number of heavy metals) and is released in the process of burning. An overwhelming amount of sulfur emissions can be traced back to coal burning to generate electricity. The table below traces the various sources and their contribution to the total emission.

Sources
coal burning utilities	74%
metal fabrication	7%
industrial boilers	6%
transportation	5%
area sources	8%

Impact of SO$_2$ on the Ecosystem

The primary impact of SO$_2$ is the creation of acid rain (an aqueous solution of sulfuric acid). The acid rain then affects the environment in a number of ways:

1. Soil and aquatic acidification

As acid rain decreases the pH of soil, the surface water pH increases. As pH increases, root structures are damaged in plants and waterborne life finds breeding an increasingly difficult task.

2. Removal of metals and minerals

The acid in the rain leaches heavy metals and minerals out of the soil. The result is that these metals (aluminum as a primary example) poison the water table and streams resulting in sterile bodies of water.

3. Species loss of both plant and aquatic life forms.

Impact on Human Health

The impact of SO$_2$ on human health is less clearly defined in terms of its impacts but the nature of the impact has been documented.

The primary health impact is increased rates of asthma attacks. Sulfur emissions take two forms, wet and dry. Wet emissions become acid rain which has few direct health impacts. Dry emissions are small particles (PM-10 or smaller) and these particles are able to evade the defenses of the body and are pulled deeply into the lungs where they damage tissue. The end result is that lung tissue becomes inflamed increasing the likelihood of asthma attacks or respiratory distress for those with chronic lung disease.

NITROGEN OXIDES (NO$_X$)

Nitrogen oxides are created by the burning of fossil fuels at high temperatures and under pressure. By themselves these compounds are harmless – 72% of the atmosphere is nitrogen. Ozone is created when NO$_X$ combine with partially burned hydrocarbons and heat (sunlight).

Ozone is a major contributor to increased asthma attacks.

> **Did You Know?**
>
> The three biggest pollutants from burning coal are carbon dioxide, sulfur dioxide, and nitrogen dioxide.

- The number of children with asthma has doubled in the last 15 years.
- 5.5 million children are affected by asthma.
- 150,000 hospitalizations each year are due to asthma.
- There was a three-fold increase in the number of deaths in children from asthma between 1977-1995.
- Over 10 million missed school days occur due to asthma each year.
- Nearly five million children under the age of 18 in the U.S. have asthma, making it the leading chronic illness for children.
- The death rate for those 19 or younger increased by 78% from 1980 to 1993 according to the National Center for Environmental Health.
- Children with asthma are 40% more likely to suffer an asthma attack on high pollution days than on days with average pollution levels according to the American Lung Association.
- Asthma deaths among children and young people increased by 118% between 1980 and 1993.
- More than 25% of the nation's children live in areas that do not meet national air quality standards. See table page 10.

The sources of NO_x are generally any burning of fossil fuels. The major sources include:

coal burning utilities	4 0 %
transportation	3 8 %
industrial boilers	1 3 %
area sources	9%

As you can see there is a very large contribution made by utilities but also by the family car. Between our cars and how we use electricity, largely at schools and homes, our collective behavior controls much of the NO_x emissions causing ozone.

GOOD OZONE AND BAD OZONE

The word "ozone" has prompted a lot of confusion over the past few years. The confusion persists because there is in fact good and bad ozone. The ozone layer in the upper atmosphere is essential because it filters harmful ultraviolet radiation from the sun, reducing the amount reaching the earth's surface. On the other hand, high accumulations of ozone in the lower atmosphere near ground level can be harmful to people, animals, crops and other materials. The ozone gas in both the upper and lower atmosphere is the same; the difference is that one benefits and one harms.

What is Ozone?

Ozone is a form of oxygen with three atoms, instead of the usual two. It is a photochemical oxidant and, at ground level, is the main component of smog. Ozone is not emitted directly into the air but is formed through chemical reactions between natural and man-made emissions of volatile organic compounds (VOCs) and nitrogen oxides in the presence of sunlight. These gaseous compounds mix like a thin soup in the ambient, or outdoor air, and when they interact with sunlight ozone is formed. Sources of these pollutants include automobiles, gas-powered motors, refineries, chemical manufacturing plants, solvents used in dry cleaners and paint shops, and wherever natural gas, gasoline, diesel fuel, kerosene and oil are combusted.

Ozone pollution is the periodic increase in the concentration of ozone in the ambient air, the natural air that surrounds us. It is mainly a daytime problem during summer months because warm temperatures play a role in its formation. When temperatures are high, sunshine is strong, and winds are weak, ozone can accumulate to unhealthy levels.

Ozone as a Health Hazard

The biggest concern with high ozone concentration is the damage it causes to human health, vegetation and to many common materials we use.

High concentrations of ozone can cause shortness of breath, coughing, wheezing, headaches, nausea, eye and throat irritation, and lung damage. People who suffer from lung diseases like bronchitis, pneumonia, emphysema, asthma and colds have even more trouble breathing when the ozone level is high. These effects can be worse in anyone who spends significant periods of time exercising or working outdoors.

Children often play outside for long periods during the summer. Their lungs are still developing, and they breathe more rapidly and inhale more air pollution per pound of body weight than adults. On days when ozone levels are high, these factors put children at an increased risk for respiratory problems.

Adults breathe more than 10,000 times each day. During exercise or strenuous work, we breathe more often and draw air more deeply into the lungs. When we exercise heavily, we may increase our intake of air by as much as 10 times our level at rest. The interaction between air pollution and exercise is so strong that health scientists typically use exercising volunteers in their research.

Materials damaged by ozone include rubber, nylon, plastics, dyes and paints. Many food crops are also damaged by ground-level ozone each year.

Did You Know?

Ozone in the upper atmosphere is called stratospheric. Ground level ozone is called tropospheric.

CARBON DIOXIDE (CO_2)

Carbon dioxide is an odorless, colorless gas that is a naturally occurring part of the environment. Specifically, the only reason life is possible on the planet is that there is a naturally occurring greenhouse effect in which CO_2 traps heat in the atmosphere stabilizing the planet's temperature and warming the planet. The problem is that the concentration of CO_2 in the atmosphere has nearly doubled in the last 100 years. As a result, the planet is likely trapping an increasing amount of energy. But what does this have to do with air pollution? Why are increased levels of CO_2 a concern?

	Pre-1998 Standards	1998 and after Standards
The Standard	125 PPB	85 PPB
How is it Measured	Highest Hourly Average	8 Hour Average (10 am to 6
Violation	When 1 hour average exceeds 125 PPB	When the 8 hour Average exceeds 85 PPB
Non-Compliance	When the third highest ozone level of a season at a particular monitor exceeds 125 PPB	When the fourth highest ozone level of a season at a particular monitor exceeds 85 PPB

The above chart demonstrates the recent changes to strengthen the health-based ozone standards. These standards have been adopted to more fully protect human health.

Many argue that increased levels of CO_2 contributes to global warming. However, the problem with increasing the amount of energy trapped in the environment has little to do with global warming; the term is misleading. The trouble with increased energy in the environment is not that the planet gets warmer (although some parts will) but rather that there will be more energy in the climate system. The result is likely to be more extremes in local weather. In other words, storms will be stronger, hurricanes will be larger, tornadoes more frequent, droughts and flash floods will be more intense. This reality has been acknowledged by the insurance industry who has to insure against these weather related problems.

Did You Know?

Central Mexico recorded a snowfall of 5" in July of 1999.

The solution to this problem is as easy to think about as it is difficult to do – reduce burning of fossil fuels. There is no known technology that removes CO_2 other than plants through photosynthesis. Given the slow speed of this solution there needs to be some form of behavior change if we are to address this problem. The change needs to be some combination of finding cleaner energy sources and using much less energy (energy efficiency) while maintaining our quality of life.

PARTICULATE MATTER

Particulate matter is made up of tiny particles in the atmosphere that can be solid or liquid (except for water or ice) and is produced by a wide variety of natural and manmade sources. Particulate matter includes dust, dirt, soot, smoke and tiny particles of pollutants. Some particles attract and combine with amounts of water so small that they do not fall to the ground as rain. Major sources of particulate pollution are factories, power plants, trash incinerators, motor vehicles, construction activity, fires, and natural windblown dust. Particles

below 10 microns in size (about seven times smaller than the width of a human hair) are more likely to travel deep in the respiratory system, and be deposited deep in the lungs where they can be trapped on membranes. If trapped, they can cause excessive growth of fibrous lung tissue, which leads to permanent injury. Children, the elderly, and people suffering from heart or lung disease are especially at risk. Particles of 10 microns or less are also referred to as PM10.

Large amounts of pollution particles in the air cause haze and can lower visibility. Particulate matter concentrations may worsen in the winter due to wood-burning and coal-burning fires that produce tiny particles of pollutants. Vehicles also emit particulate matter, which can cause higher pollution levels in more densely populated areas. Highs or lows may also be caused by area-wide weather conditions such as dust storms or rain. Some areas within a city may be worse than others if they are located closer to major pollution sources such as industry.

Government Regulation

There are three tools that government can use when trying to reduce air pollution. First, government can **regulate** the behavior by making it illegal. Second, government can offer **incentives** in order to make industry (or individuals) want to change their behavior. The third option is to create a **choice** for industry by telling them what has to be done and how long they have, but allowing the industry to decide how it is to be done. This third option is a compromise between the first two tools. In different situations, some of these tools work better than others.

REGULATION

It is the responsibility of the government to protect all of a country's citizens equally. In order to carry out this responsibility, the government must pass laws in order to ensure that some individuals do not violate the rights of other individuals. Usually this is necessary when one citizen has more power due to money, influence or ability. Similarly, there are certain behaviors which society as a whole has deemed undesirable - such as murder, abuse, and robbery. However, there are other behaviors, such as polluting whose negative effects are not as evident as others. It is in this gray area that the government has the responsibility to monitor activity and to pass laws which equally protect all citizens.

In order for regulation to work, there needs to be a very specific type of behavior which must be changed. One of the strengths of regulation is that it is tightly focused on specific problems. However, this can also be a weakness since tight regulation can exclude other similar behaviors that are not perceived to be as problematic. The more complex the issue, the more difficult it is to define the problem into an easily solved package.

In addition to the issue of complexity, issues where regulation works well have a clear, limited solution available to the problem. For this to exist, there needs to be a solution, a limited number of actors and a means of monitoring and enforcing compliance. When all of these factors are in place, regulation is a powerful tool for changing behavior and addressing pollution problems. However, when one of these elements is missing, regulation become ambiguous. If there are not clear solutions, if the number of individuals causing the problem is unknown, or if there is no way to monitor whether or not the regulation is being followed, laws and regulation are useless as a tool to control air pollution.

Did You Know?

One ton of SO2 released into the environment causes $4000 in damage.

INCENTIVES

The second tool, incentive can be viewed as the carrot to the regulatory stick. In an incentive based system, the state selects a type of behavior they wish to promote and then tie that action to some type of reward. An example of this was when the Environmental Protection Agency (EPA) determined that a major source of water pollution was leaking septic tanks. The EPA offered to pay 95% of the cost of build sewer systems. In this case the community could offer improved services and the environmental goal was met. In effect it was a win-win situation.

The trouble with this strategy is that it is very expensive. The cost associated with providing the incentives tends to be quite high. The overall costs can only be justified when the issue in question is almost fully correctable by the action being paid for. Otherwise

it is hard to get Congress to pay that kind of bill. As the governmental budget tightens, incentive programs become increasingly difficult to justify without an absolute guarantee of success. As the number of cases where this degree of certainty exists is very limited this is a rarely used option.

CHOICE
(Market-Based Tools)

A middle ground between the carrot of incentives and the stick of regulation is the use of market tools. The entire category of market tools is a set of policy creations that begin with the assumption that the forces of the market (the invisible hand) can be used to reduce pollution in a low cost, automatic way. The idea is that regulation is used to establish a series of pollution emission goals for polluters. In addition there needs to be some combination of rewards and punishments for meeting or failing the target reductions. Once this is done the polluter is free to make a series of decisions - how best to meet the goals, when should those goals be met, and is the economic rationale powerful enough to meet those goals at this time. Finally, there needs to be very specific data on just how much pollution is produced by a given factor, how it is produced and what are the opportunities for change.

One example of a market tool is the creation of the Sulfur Dioxide Market to address the acid rain problem. In this case the EPA set annual SO_2 reduction targets for each major coal burning utility in the country. They then awarded allowances, representing the right to emit one ton of SO_2, based on each utility's historic emissions and reduced that amount annually based on the annual target. These utilities had to give back the allowances equal to the total emissions for that year. Any extra allowances could then be traded on the Chicago Board of Trade. The incentive to meet the reduction target is that if you were cleaner than required you could sell your extra allowances and if you failed to meet the target then you had to buy allowances. This means that cleaner utilities are rewarded with a new source of income (allowance sales) and that dirty utilities have to pay their cleaner competitors for the allowances they need to meet the EPA requirement so that they have an allowance for each ton of pollution.

The outcome of this market has been impressive. First, the total amount of SO_2 emitted has decreased by over 40% in the past six years. Second, the market has driven much of this change with no real complaint from the utilities – they can understand this system and make economic decisions that meet their best interest. Finally, as a result of this calculated decision of when to invest and when to buy allowances and delay, the cost per ton removed has dropped sharply from the price of $1,200/ton under the previous regulatory regime.

History of The Clean Air Conservancy

The Clean Air Conservancy is the only non-profit, non-partisan, market-based environmental organiza-tion committed to reducing air pollution. We participate in pollution markets on a continual basis. Our organization has built a national reputation as the leader in this innovative approach to improving and addressing environmental quality. Our method of reducing pollution is simple, direct and effective: We acquire marketable pollution allowances and retire them forever—preventing tons of sulfur dioxide from ever polluting the environment.

The Clean Air Conservancy began in 1992. Our vision is to prevent air pollution by affecting systematic changes in the economy; specifically changes in the behavior of polluting entities. We also aim to encourage and facilitate greater public participation in preventing air pollution through effective environmental tools.

The tools to accomplish these goals came through the Clean Air Act Amendments of 1990 that introduced a nationwide market system to cap utility sulfur dioxide emissions. This system is a radical departure from command and control pollution regulation since it introduces economic incentives for companies to reduce their emissions. Utilities now have the incentive to sell allowances that permit them to emit sulfur dioxide if they reduce their emissions below their EPA allotment. In this way, the market system encourages participants to incorporate the cost of emitting pollutants into their production costs, thus putting a real price on polluting.

Did You Know?

Leaving an incandescent bulb on for one hour burns over 100 pounds of coal.

As a result, emissions have been reduced more than initially expected.

We are pleased to present a very positive and encouraging history of success. Since 1992, we have retired over three million pounds of sulfur dioxide (SO_2). The Environmental Protection Agency has put a value of six million on the environmental and health damage that was prevented in relation to the SO_2 retired.

During 1996, we had a five-fold increase in the number of schools engaged in our retirement and curriculum programs. We further developed a school energy audit that lets students assess the damage their classroom contributes and helps them associate a number with the amount of energy their classrooms waste. To date, we foster a network of 130 schools that extend from California to Maine to Guam and that number continues to grow.

The Clean Air Conservancy has the organizational capacity and the vision to provide the public with information about, and access to, participation in effective pollution reducing markets. Our staff educates the public about the causes and effects of air pollution and actively ensures that the public interest is served by participating in the creation of new market-based approaches to pollution control. Perhaps most importantly, we facilitate the retirement of pollution allowances by private sources through direct donations to the CAC.

WHERE WE ARE HEADED

The future is bright and broad for the pollution regulation market. Recent EPA guidelines make pollution a more important issue to businesses and individuals alike. The demands for a cleaner environment and greater accountability for pollution have made their way from small towns across the country to decision-makers in Washington. Regulating pollution can be a point of contention between regulators and industry or it can be a way to achieve economic prosperity and a cleaner environment simultaneously. We see the pollution market as the best way to satisfy the public and industry.

Our plans to move ahead and grow our organization in size and scope include moving into a few different directions. Specifically, we need to grow school involvement in the pollution market. One hundred and thirty schools is a wonderful start, but it is hardly our end goal. Children are the future; helping them foster a positive view of how business and environmentalism can work together is what we want our school program to accomplish.

We are being even more active in the creation of new pollution markets. Instead of being reactive to EPA guidelines, we hope to become proactive in the development and implementation of new pollution markets. Sulfur dioxide is not the only contributing agent to air pollution. Locally, it is our plan to outreach to business, political and environmental communities with the goal of creating an ozone market to help the region comply with tougher ozone standards. Our single largest challenge may well be finding the funding to provide staff time to carry out this massive initiative. While these initiatives are rather narrowly focused and connected in their infancy to the local area, the problem of pollution does not plague the United States alone.

The pollution market will eventually go international. The United Nations and the European Union are studying the logistics of an international market currently. We want to be the "expert" when the market is developed. As the single environmental organization that uses the market approach to prevent pollution, we have a jump on every other potential group. We understand the market, the pollution and the actors. Since we do not litigate, we are viewed as an "honest broker" in the market.

Control Technologies

Burning coal has been the primary source for generating electricity in this country for the past century. Coal is an attractive source of energy because it is cheap, abundant, and accessible. Coal technologies have been developed and improved over the past 25 years to meet the nation's high demand for electricity. According to the Energy Information Administration (EIA), coal was used to generate over fifty percent of our electricity in 1995. This trend is expected to continue through 2015.

In order for coal to continue to respond to the United State's electricity needs as a cheap and reliable resource, it is imperative that it be able to meet new and emerging environmental standards. In addition, due to recent electric deregulation, coal-burning utilities must be able to operate in an economically competitive manner. Pollution control technologies for fossil fuel burning utilities have been developed in order to accommodate both of these requirements. Below is an explanation of some of the technologies for controlling nitrogen oxide, particulate matter, and sulfur dioxide along with a brief description of the environmental regulation that makes them necessary.

NITROGEN OXIDES

Nitrogen oxide (NO and NO_2, collectively referred to as NO_X) emissions are generated primarily from transportation, utility, and other industrial sources. NO_X is a contributing factor to acid rain, ground level ozone, and visibility degradation. Title I of the Clean Air Act Amendments of 1990 (CAAA) includes new requirements for NO_X as a contributor to the national problem of ground level ozone. It requires:

- existing major stationary sources to apply reasonably available control technology(RACT)
- new or modified major stationary sources to offset new emissions and install controls representing the lowest achievable emission rate
- every state with an ozone non-attainment region to develop a State Implementation Plan(SIP) that, in most cases, will call for reductions in stationary source NO_X emissions well

Did You Know?

Since starting, The Clean Air Conservancy has retired enough SO_2 to prevent $5,000,000 in damage to the environment.

beyond those of the RACT.

In addition, Title IV of the CAAA (also known as the Acid Rain Program) addresses controls for specific types of boilers, including stationary coal-fired power plants. Phase I of this program became effective in 1995 and required two types of boiler systems (tangentially fired and dry-bottom wall fired) to reduce their emissions. Phase II of Title IV became effective in the year 2000. Under these regulations, all fossil fuel fired units are forced to comply with lower emission limits.

In response to the new NO_X standards and the potential for more stringent limitations in the future, control technologies are being tested and improved to meet these regulations.

NO_X control can be achieved through modifying the combustion process, acting upon the products of combustion, or a combination of both.

Low-NoX Combustion (LNB)

NO_X is formed from the oxidation of the nitrogen bound in the coal (fuel-NO_X) and the oxidation of nitrogen in the air at high temperatures of combustion (thermal NO_X). In order to mitigate the fuel-NO_X it is important to limit the amount of oxygen at early stages of the combustion process. To control thermal-NO_X it is important to limit peak temperatures. Low NO_X combustion or burning (LNB) is a combustion control. It reduces the formation of NO_X by maintaining a primary combustion zone that is deficient in oxygen and by completing the combustion process in stages. This staging of combustion controls the rate at which

air needed for combustion is mixed, avoiding a highly oxidized environment and hot-spots conducive to the formation of both types of NO_X. LNB is the primary technology currently used for NO_X reduction.

LNB combined with the principle of over-fire air (OFA) can further reduce NO_X formation. OFA limits the production of additional NO_X by diverting a portion of the oxygen needed for complete combustion from the burners to an upper furnace. This causes combustion to occur with less oxygen than required, slowing down the formation of fuel-NO_X.

In addition, reburning can be used to decompose NO_X once it is formed. The majority of the coal is burned in the primary combustion zone. When additional fuel (coal, oil, or natural gas) is injected, further burning occurs in a fuel-rich, oxygen deficient zone higher in the boiler. NO_X entering this zone is stripped of oxygen, forming elemental nitrogen, which can be released to the atmosphere as a benign product.

Technology	NO_X Reduction %
LNB	30-55
LNB + OFA	60-67
Reburning	50-70

Selective Catalytic Reduction (SCR)

SCR is a post-combustion control. Although LNB is the most common control technology, as more stringent requirements are enforced, utilities may be forced to use post-combustion technologies such as SCR, either alone or in combination with LNBs.

SCR is a technology installed downstream of a power plant boiler. It uses

a chemical reaction to break down the NO_X present in the flue gas released after combustion. Ammonia (NH_3) is injected into boiler flue gas. The flue gas is then passed through a catalyst bed where the NO_X and NH_3 react to form nitrogen and water vapor in the presence of oxygen. SCR catalysts are made of a ceramic material that is a mixture of carrier (titanium oxide) and active components (oxides of vanadium and, in some cases, tungsten).

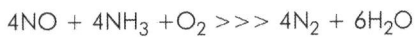

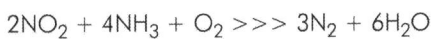

$$4NO + 4NH_3 + O_2 >>> 4N_2 + 6H_2O$$

$$2NO_2 + 4NH_3 + O_2 >>> 3N_2 + 6H_2O$$

The harmless nitrogen gas can then be released into the atmosphere.

SCR has the potential to reduce NO_X emissions by 80-90%. It is one of the few technologies capable of removing these high levels from high-sulfur coal. SCR is applicable to all types of boilers and can be used with new and existing power plants. However, SCR is a more expensive option relative to LNB and SNCR.

SELECTIVE NON-CATALYTIC REDUCTION (SNCR)

SNCR is similar to SCR in that it uses a nitrogen-based reagent, most commonly urea (CH_2CONH_2) or ammonia (NH_3), injection in the flue gas to convert NO_X emissions to elemental nitrogen and water. The key difference is the absence of a catalyst because the system is operating at a much higher temperature.

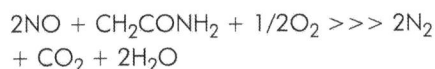

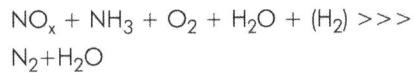

$$2NO + CH_2CONH_2 + 1/2O_2 >>> 2N_2 + CO_2 + 2H_2O$$

$$NO_X + NH_3 + O_2 + H_2O + (H_2) >>> N_2 + H_2O$$

SNCR technologies are considerably cheaper than SCR because it does not require a catalyst. However, SNCR NO_X reductions are 35-50% compared to the 80-90% of SCR. In addition, SNCR technologies are limited to smaller boilers.

PARTICULATE MATTER

Particulate matter is one of the six criteria pollutants affected by the National Ambient Air Quality Standards (NAAQS) of Title I in the CAAA of 1990, which sets limits for particulate emissions.

Bagfilter/Baghouse

Baghouses have been used extensively during the past decade because they are efficient at gathering dust. The particles and flue gas are separated in tube-shaped filter bags arranged in parallel flow paths. The particles are collected either on the outside (dirty gas flow from the outside-to-inside) or the inside (dirty gas flow from inside-to-outside) of the bag.

Baghouse technologies are commercially available throughout the world. Baghouses have demonstrated a collection efficiency of 99% for particles in the 0.05-1.00 micron range. However, they too have a high capital cost.

Did You Know?

The amount of pollution emitted by a power plant depends on when the coal was mined.

Electrostatic Precipitator

Electrostatic precipitators are large industrial units designed to collect dust particles. The precipitator charges particulate matter by static electricity. These charged particles are then attracted to and deposited on collection areas. The collection areas are shaken and the dust is collected in hoppers for disposal.

Precipitators are used to reduce environmental air pollution and to collect chemical particles for another use. They are capable of collecting 98-99% of particles from the gas stream.

SULFUR DIOXIDE

Sulfur Dioxide (SO_2) is the leading contributor to acid rain. The problem of acid rain is addressed in Title IV of the CAAA, which sets up a SO_2 pollution allowance trading system. Coal burning utilities are given a set number of allowances. Each allowance represents one ton of SO_2. These allowances may be sold and traded. Each year, the number of allowances distributed is reduced, forcing the price per ton to rise. This program encourages utilities to install pollution reducing technology, since the cost to pollute is increasing and availability of allowances decreases annually.

Phase I of this program began in 1995, creating a market-based system for pollution reduction that applied to 435 utilities across the country. The requirements for Phase I were largely met by a switch to low-sulfur fuel. Phase II went into effect in 2000, and increased the number of utilities subject to lower pollution standards to over 2000. Since it is unlikely that these requirements can be met by fuel switching and allowance purchasing alone, utilities will be forced to install their choice of a SO_2 control technology.

Dry Flue Gas Desulfurization (FGD)

The process of dry scrubbing involves the introduction of a drying gas (lime and water mixture) to the flue gas, which mixes with the SO_2 to form a dry product that can be separated in the particulate removal equipment (baghouse or ESP) and sent to a disposal site.

The dry scrubber has a removal efficiency of 70-90% with low sulfur coals. Preliminary testing demonstrates that similar efficiency with high sulfur coals is also achievable. The dry scrubbing process requires less power to complete than wet scrubbing. Dry scrubbers have a lower capital cost than wet scrubbers and are easier to operate and maintain. However, as wet scrubbers are being simplified and costs reduced, it is increasingly difficult for dry scrubbers to complete on cost-effectiveness (cost per ton of SO_2).

Wet Flue Gas Desulfurization (FGD)

In wet scrubbers, the flue gas enters a large vessel where it is sprayed with a calcium-based water slurry (limestone, lime or sodium hydroxide). The calcium in the slurry reacts with the SO_2 to form calcium sulfite or calcium sulfate, which are removed by dewatering and settling into a thickener. These solid wastes are then collected in the baghouse or ESP and disposed of in a landfill. Some of the byproducts can be marketed as building materials.

$$CaCO_3 + SO_2 + H2O + O_2 >>>$$
$$CaSO_3 + CaSO_4 + CO_2 + H_2O$$

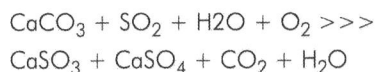

Wet scrubbers are usually designed for efficiency of 80-90% SO_2 removal for both high and low sulfur coals. Wet scrubber technology is the most proven and commercially established SO_2 removal process in most developed countries. In the last decade, the cost of wet scrubbing has been significantly reduced.

Renewable Energy

Unlike fossil fuels, renewable energy is an energy source that is naturally replenished and can be converted to human use. In order to earn the renewable energy label the energy source has to be naturally occurring (wind, flowing water, etc.) and replenish itself without human intervention. The key is that it naturally **renews** itself as opposed to fossil fuels which are naturally formed, but once used they are gone.

Using these renewable energies to generate electricity has another benefit: renewable energy tends to generate no pollution. In other words, renewable energy replenishes itself, has little, or no pollution outcomes and there is some renewable form available in almost every location on the planet. The only negative is typically a high cost to convert the energy into a form we can use. For example, there is more energy in wind than we could ever use but in order to make it commercially viable we need to build a turbine and get a high enough wind speed to generate meaningful amounts of electricity. Fossil fuels are much easier to use.

In order to understand renewables we will look at each of the main types of renewable energy. We will begin with a discussion of coal as the most commonly used fossil fuel in order to place renewable sources in context.

COAL
Coal is the final step of a long process in which sunlight is converted into chemical energy. After collecting energy from the sun through photosynthesis, plants die, decompose and are compressed by the weight of successive generations of plant material. Over time this matter becomes coal and its formation also incorporates the materials surrounding it, often toxic compounds (mercury, sulfur, uranium, cadmium, and lead). Coal is then easily mined and burned, converting the stored energy into hot water. This hot water can then be used in heating and cooling buildings.

SOLAR POWER
Solar power is the direct capture of energy from the sun and its conversion into usable power. Solar power has two types. The first form of solar power is passive, where the sun's heat energy is focused in a collection and transferred into hot water. This hot water can then be used in heating and cooling buildings. The second form of solar power is generated by sunlight in a photovoltaic (pv) panel, which converts sun energy into electric current. The pv panel generates direct current which can then be used to operate any electric equipment. The good thing about pv panels is that they can operate almost anywhere and they generate pollution free energy. The downside is that solar energy still cost almost twice the

price of coal generated electricity, despite the cost being reduced by 50% over the past 10 years. However, the price of the pollution generated by coal burning is not calculated in these numbers. It naturally follows to ask "What is clean electricity worth to you?"

WIND

Wind power is one of the oldest means of transforming energy into work. The Netherlands owes its continued existence to wind power pumping water out from behind dikes. Netherlanders have used wind to pump water for over 200 years. This form of wind power was very inefficient because winds die down; but without electric alternatives that was never a concern.

We now have a whole new technology that generates electricity by using the turning blades to spin a generator which produces electric current – much the same way a hydroelectric facility uses falling water. This newer technology has evolved over the past twenty years, reducing the cost/kwh from $.20/ kwh to about $.07/kwh. The energy is emission free with no fuel costs.

The limitations to this form of power are twofold. First, the power generated is a function of wind speed. To be cost competitive with other forms of power there needs to be winds of 11 to 12 miles an hour, which does limit the areas where wind power is likely to be effective. Secondly, the power is only generated when the wind is blowing so unless you have the capacity to store a lot of energy (for example, in a battery), wind power is too variable to be your only electricity source.

HYDROPOWER

Hydroelectric power is perhaps the oldest means humans found to harness natural occurring power. Whether through the use of a water wheel to drive gears, or a turbine built into a dam to generate electricity, we have been using water power since ancient times. At its core hydropower is very simple: we use the momentum of moving water to turn a shaft which in turn then does some type of work. The only questions that differentiate different types of hydropower is how much momentum the water has and how effectively you can capture that energy and convert it to work.

Modern hydropower comes in two major types, run of the river and impoundment of water. "Run of the river" places generators on low barriers in the river (locks, levies etc) and allows the normal flow of the river to turn the turbines. While this generates relatively small amounts of energy (low levels of momentum) it has minimal construction costs and has minimal impact on the river ecosystem. It can be used on most rivers and streams.

Larger scale hydropower is based on the idea of impounding large amounts of water behind a dam and then allowing that water to fall a great distance (50 ft. or more), increasing its momentum, thus driving the turbine. As the height of the fall increases, so does the momentum which increases the amount of power generated. This technology generates a great deal of power and does so at prices competitive with, or lower than, coal. There are no air emissions associated with this technology; however, hydropower does have some limitations. The number of rivers with exploitable sites for power dams is very limited. Also, dams totally transform the river behind them and are especially hard on migratory aquatic life – they can not get around a 150' dam.

WASTE GAS

This is a very new technology. In waste gas recovery, landfills are mined for pockets of methane gas that result from the decomposition of food, yard wastes and other organics – much the way natural gas formed. Methane burns hot and cleanly with only CO_2 as a byproduct. It can be used to drive a turbine to generate electricity like any gas burning plant.

The positive side is that methane is a very powerful greenhouse gas and by converting it to CO_2 it actually reduces the impact on global climate change. Further, methane is highly explosive and its removal and combustion decreases the chances of explosion and landfill fires. The downside of waste gas is that it can be difficult to find and tap.

FUEL CELL

Fuel cell technology is a new and intriguing technology. In effect, a fuel cell generates electricity and its sole emission is water vapor. Even better, there are no moving parts to break and it has the highest energy conversion rate (the rate at which one type of energy is converted into electricity). In effect, a fuel cell converts 70% of the energy put into it into electricity. Fuel cells are also small, silent, and able to generate almost unlimited power in your basement.

A fuel cell works by passing hydrogen gas through a semi-permiable membrane. The passage of the hydrogen atoms through the membrane displaces electrons which then produce an electric current. This entire process is clean and can be paired with other technologies, such as wind power in order to make the fuel cell a battery to store energy. For example, the windmill generates electricity when there is wind and that electricity is used to split water into hydrogen which is stored in a tank. The hydrogen is then used to generate electricity through the fuel cell whenever it is needed. In effect, this becomes a low cost means of storing variable wind power in a chemical form (hydrogen) and then using it to generate power when it is demanded at a specific location. For more information about fuel cells, check out www.alternativepower.com.

It is important to understand the different ways that electricity can be generated. As this country moves toward electricity deregulation, individuals will be given a choice of where to buy electricity. One type of electricity which will be marketed is "green" energy. Knowing how energy is generated is important if one is to make informed decisions. Also, with some of the new technologies, individuals will be able to generate electricity at home and sell the "extra" back to utility companies. The key questions to remember are "How green?" and "What are we trading off for cheap coal power versus clean power?"

Topic One:
The Different Pollutants and Their Sources

SKILLS:

Scientific Procedure
Environmental Issues (Air Pollution, Acid Rain, Ozone)

PURPOSE:

To introduce different types of air pollution/air pollutants to the students and to be able to recognize their sources.

OBJECTIVE:

To begin the unit, you need to introduce the students to the existence of air pollution in the world around them. They will all participate in the experiment and should have some understanding of what they will expect to discover. They should be able to explain some of the major causes of air pollution in the world around them.

FOCUS:

In order to provide a foundation for further lessons which relate to environmental pollution issues, it is extremely important to provide the background to the students. The first thing that the teacher needs to do is introduce the concept of air pollutants. In order to facilitate this discussion, you may want to conduct the Visible and Invisible Air Pollutants Experiment. Be sure that you discuss with the students what type of results you expect. Let the students come up with ideas as to where the jars should be placed. Ask them what they expect to find in each different area. Through the completion of that activity students will become aware of pollution in their school environment.

PROCEDURE:

1. Begin a discussion which focuses on the pollution with which the students are familiar.
2. Introduce some of the specifics about CO_2, NO_x, and SO_2. Discuss sources of these pollutants.
3. Conduct some of the experiments which might demonstrate the different pollutants.

Topic Two:
Environmental Effects of Air Pollution

SKILLS GAINED:
Scientific Procedure
Observation
Chemical Reactions

PURPOSE:
To understand the harmful effects that the air pollutants have on the environment

OBJECTIVE:
After presenting to students the different types of pollutants, it is important to understand how these different pollutants affect the environment. By discussing some of the various adverse effects of sulfur dioxide, nitrogen oxide, carbon dioxide and particulate matter, students begin linking the air pollution with the environmental effects. Through discussion and experiments, students can begin to see the real world ramifications of air pollution.

FOCUS:
If you have not already done so, introduce the students to some of the different air pollutants. Discuss the environmental and health effects of each. Make sure that students can link the specific pollution with its environmental effects.

PROCEDURE:
1. Begin a discussion of pollutants and link each with its health and environmental effects. You might want to also add some discussion about the source of these pollutants.
2. Do some of the simple experiments which can demonstrate the effects of acid rain, carbon dioxide and ozone. Many of these experiments have immediate results which you should discuss with the students. Other experiments or projects take more time. Set these up and review with the students the timetable for completion.
3. Afterward lead a discussion which ties the experiments to the pollutants.

Topic Three:
Government Regulation:
Options for Limiting Air Pollution

SKILLS GAINED:

Citizen Action
Government Bureaucracy
Economics/Markets
Legislative History

PURPOSE:

To understand the different types of regulation for industrial pollution, recognize what works, what does not and why, and focus on the current market solution to pollution control.

OBJECTIVE:

Upon completion of this topic, students will have a clearer understanding of the different ways the government tries to control industrial pollution. The EPA will be introduced as will the Clean Air Conservancy. Students will be given a glimpse of the Conservancy's role in limiting air pollution.

FOCUS:

Through discussion and the activity, students should come to an understanding of pollution regulation and the history of the EPA's control of air pollution. The game will also provide students with a concrete example of a market.

PROCEDURE:

1. Discuss some of the history of regulation (as much as you believe your students will understand or as much as you deem necessary to convey t h i s point). While not necessarily emphasizing market solutions, explain that this is one of the most beneficial forms of regulation currently available.
2. Playing "The Allowance Game" is a good way to reinforce some of the concepts of a market driven regulation. Before playing the game, make sure to go over the information which will provide much of the needed background for understanding the game and its rules. Ensuring that they have an understanding of the way the market works will help them in playing the game.
   a) During the game, you might offer help during open markets in getting the ball rolling, but mostly you should remain as an observer to the dynamics being created.
   b) Emphasize the Conservancy's role during the game. This will establish a foundation for the students in understanding how effective citizen participation can be in controlling pollution.
   c) The students should be discovering many of the principles of the market (profit, demand, etc.).

d) At the end of the game, conduct a discussion about some of the things learned while playing. (It might work best to make an overhead of the game board and play as a class divided into groups.)

Topic Four:
Alternative Ways to Control Air Pollution:
Energy Efficiency, Alternative Electricity Generation

SKILLS:
Civic Responsibility
Word Problems
Internet and Library Research
Cost Comparison and Analysis
Alternative Energy Sources

PURPOSE:
Students will gain an understanding of the relationship between responsible energy usage, energy conservation and reduction of air pollution. Students will also become familiar with different ways electricity can be generated.

OBJECTIVE:
Through a discussion about energy efficient technologies (lights, computers, windows, insulation, etc.), an audit of their classroom, and research on the cost benefits of energy efficiency students will be able to recognize ways in which to control pollution and save money simultaneously. Further, students will become familiar with alternative solutions for energy generation (solar power, wind power, hydropower).

FOCUS:
Focus on the link between electricity generation and pollution emissions. Establishing and understanding this connection is vital in recognizing one of the simplest ways, as an individual, to reduce air pollution. Also, enforce the idea that electricity is not "bad" but that burning coal to produce electricity causes air pollution. Emphasize the "clean" technologies which exist to eliminate air pollution.

PROCEDURE:
1. Lead a discussion which links electricity generation and pollution emission. Ask questions such as "Where does air pollution come from?" and "What are different ways to generate electricity?"
2. Discuss with the students different ways that they might decrease the amount of pollution they use at school and at home.
3. Through the Energy Audit determine the amount of electricity and pollution your classroom generates during the school year.
4. After completion of the Audit, discuss ways that you can save electricity (thus reducing pollution) in your classroom. For example, suggest turning off computers, lights, air conditioners, etc. when they are not being used.
5. Do the **Energy Efficiency** research project to determine how much elec-

tricity, money and pollution could be saved by making simple changes in the classroom.

6. Lead a discussion about different ways to generate electricity.

Topic Five:
Using Technology to Control Pollution

SKILLS:
Scientific Method, Civic Responsibility, and Simple Principles of Electricity

PURPOSE:
To discuss ways to control the amount of pollution emitted by electric utilities. To find ways to limit the amount of electricity used in everyday lives.

OBJECTIVE:
Upon completion of these activities, students will have an understanding of the role of electricity and its relationship to air pollution. Students will also have an understanding of what can be done not only to conserve electricity to control air pollution, but also some of the ways that pollution can be cleaned before it enters the air.

FOCUS:
An ideal link to this topic would be some of the basic principles of electricity — how it is generated, etc. This would help to reinforce the relationship between the generation of electricity and air pollution. Explain to the students that electricity is a vital part of their every day lives, but at the same time electricity contributes to air pollution. If you have done the Energy Audit with your students, you can use some of the information gained through that activity to illustrate how much electricity is needed to run even a simple classroom. Making the wet scrubber gives the students a clearer picture of some of the options available to utility companies for cleaning the air.

MATERIALS:
1. Idea sheet for reducing the amount of electricity you use
2. Directions and materials to build a wetscrubber
3. Pad of paper or chalkboard to facilitate discussion
4. Dr. Bob Video

PROCEDURE:
1. Lead a discussion about the importance of electricity in our lives.
2. Show the video which will explain generation of electricity and how control technologies can be used to control pollution emissions.
3. Discuss the different ways that utilities can control pollution.
4. Follow the procedure for creating the simulated wet scrubber.

ENRICHMENT:
In the appendix, there is a procedure for making an electrostatic precipitator. Have a group of students conduct this experiment for the class and explain the principles involved. Also have some students do further research on the different types of ways utility companies might control pollution in their necessary emissions.

Topic Six:
Conclusion

PURPOSE:
To provide students with an opportunity to demonstrate their understanding of the topics presented

OBJECTIVE:
To assess students understanding and ability to incorporate the topics presented

FOCUS:
The conclusion provides the student an opportunity to draw together much of what was learned as well as explore topics not specifically or thoroughly covered in this unit. Some suggestions for research projects are included in the curriculum unit, but are by no means limiting. Teachers should explore student interest based on the feedback and ability of his/her students.

PROCEDURE:
Through whatever means you feel most comfortable, assess the students understanding of the topics. Included in the curriculum unit is an outline for research projects as well as a play which students can use to share their new knowledge with others. These topics are by no means limited to the ones we established. Also, you need to remember to conclude all of the long-term experiments.

Teacher Preparation

TOPIC ONE

Preparation Time: 15 minutes

1. Familiarize yourself with the different types of pollution (NO_x, SO_2, and CO_2).
2. Gather all the materials necessary for conducting the Visible/Invisible experiment. If you need additional information about any of the types of pollution, contact the Clean Air Conservancy and we will be more than happy to help.
3. Check out the lab reports online for the different experiments.
4. Make enough copies of the worksheet for Topic One for all of your students. (Page 99)

TOPIC TWO

Preparation Time: 15 minutes

1. Familiarize yourself with the effects the different pollutants have on the ecosystem and on individual health.
2. Gather all the materials necessary for the different experiments (Carbon Dioxide, Harmful Effects of Acid Rain, The Rubber Band Air Test, Temperature Inversion 1& 2) that will be used to further demonstrate the concepts of pollution effects.
3. Check out the lab reports online for the different experiments.
4. Make enough copies of the worksheet for Topic Two for all of your students. (Page 100)

Topic Three

Preparation Time: 30 minutes

1. Read about regulation and markets to have background for this topic.
2. Copy enough of the Allowance Game Information sheets for all of your students. (Pages 91-92)

3. Preview the Clean Air Conservancy video to decide if you are going to use it with your students.
4. Make enough copies of the game board that each group will have one or make one copy on a transparency if you are going to play together as a class. (Page 38)
5. Copy the money, pollution cards, chance cards, and allowance cards so that there are enough for each group. Make as many copies as indicated for each group. (If you are playing as a class, you will only need one set. You also might want to laminate some of the cards and the game boards so that they can be used again in following years.) For easier playing, copy the different types of cards and money onto different colored pieces of paper. This will make it easier for students during the game.
6. Cut out all the different pieces.
7. Gather markers and dice to play the game.
8. Make enough copies of the worksheet for Topic Three for all of your students. (Page 102)

TOPIC FOUR

Preparation Time: 20 minutes

1. Make copies of the Energy Audit Information sheet for your students (Page 93). Possibly hand them out as homework the night before you are to do the audit.
2. Make enough copies of the Energy Audit Worksheet for your students.
3. Do the Energy Audit before class so that you can anticipate any problems students might have.
4. Each student will need a calculator.
5. In order to complete the energy audit, you will need information from the school's maintenance

director or custodian. You either need to ask ahead of time so that you can provide the students with the necessary information, or you might even want to ask the maintenance director to visit the classroom on the day while you do the audit so that he/she is available to answer whatever questions the students might have. You probably want to provide this person with a copy of the audit and audit information sheetbefore they come to your class-room so that he/she has the answers necessary or can direct you and the students to the person who might have the necessary answers. Some questions you might want to ask:

- What kind of insulation does our school use?
- How many inches of insulation are in the walls?
- What kind of windows do we have?

6. Make copies of the Renewable Energy Project.
7. Make enough copies of the list of ways individuals can limit the amount of electricity they use for all of your students.
8. Make enough copies of the worksheet for Topic Four for all of your students.

TOPIC FIVE

Preparation Time: 20 minutes

1. Read about the different Control Technologies.
2. Preview the "Dr. Bob" video on control technologies.
3. Gather together the necessary materials for making the wet scrubber.
4. Make an overhead copy of Figure 1 so that all the students can see what

you are trying to do.
5. Set up the Electrostatic Precipitator Experiment or make copies of the procedure to assign the task to students.

TOPIC SIX

Preparation Time: 20 minutes

1. Examine the conclusion projects and decide which one(s) apply to your lesson plan.
2. Make copies of the necessary materials for any of these projects.

The Allowance Game

THE RULES

The first step in playing this game is reading the information sheet about Pollution Markets. If you have not done so, please read this before you begin playing.

Focus:

Throughout the game, you will produce and reduce pollution and have the opportunity to buy and sell allowances with your competition.

Objective:

You need to have at least one allowance card per pollution card when the game is over. The player with the most money at the end of the game is the winner. (Even though the environment is always the true winner).

There are 21 spaces on the game board - Start, four Open Market spaces, five Chance card spaces, eight Gain and Lose Pollution spaces, and three "Purchase a Control Technology" spaces. The start space is also a "Purchase Control Technology" spot. Each space directs the player to perform a different task. Follow the directions on the square in order to proceed.

One player is designated as the EPA representative (Note: Although this person works with the EPA, he/she still has his/her own factory which emits pollution). The purpose of this individual is to pass out money and allowances as directed throughout the game. If there are allowance cards available in the EPA "bank", the EPA representative may sell these pollution allowances for $5000 each to a player before his/her turn if that player wishes to buy some. The money goes to the bank, not the EPA representative.

Start:

Here is where everyone begins. Before you roll, each player receives 1-$5000, 5-$1000, 8-$500, 10-$100 ($15,000), 10 Allowances cards, and 10 Pollution cards. Each player places his/her marker on the start square. Each player will then roll the die to determine who goes first. Whoever rolls the highest number, rolls again and then moves his/her marker forward the appropriate number of spaces.

Passing Start: Every time you pass the start square, your customers pay their bills. Every player collects $2000 when they pass start. However if you have a control technology, you also lose pollution cards every time you pass start. If you land exactly on the start square, you may purchase the control technology of your choice. Make sure you properly deduct pollution cards since you now have a control technology. For example, if Rob lands on start and purchases a SCR technology, then he will gain his $2000 AND lose 3 pollution cards.

Follow the list below to determine how many pollution cards you will lose:

Technology	Pollution Reduction	Cost
BAG HOUSE	5 Pollution Cards	$5000
SCR (SELECTIVE CATALYTIC REDUCTION)	3 Pollution Cards	$3000
WET SCRUBBER	2 Pollution Cards	$2000
REBURNING	1 Pollution Card	$1000

Open Market

1. Open market means an opportunity to buy and sell allowances from the other utility companies (the other players). When a player lands on the Open Market square, all players may buy and sell pollution allowances from each other.
2. Besides the players buying and selling pollution allowances, the Clean Air Conservancy has an opportunity to buy allowances during an Open Market. The player who landed on the Open Market square rolls the die to determine how many allowances the Conservancy purchases. If the player rolls a 1 or 2, each player must forfeit one allowance to be retired from the game. Once forfeited, these allowances can not be purchased from the EPA representative or any other player for the rest of the game. If the player rolls a 3, 4, 5, or 6, then the Clean Air Conservancy does not retire any allowances during that Open Market.
3. At no time is any player obligated to buy or sell any allowances. The owner determines the price of each allowance. Remember, you must have at least one allowance card for each pollution card.
4. It is important to keep track of the value of your allowances. At the end of each Open Market, average the price paid per allowance. (So if three allowances were sold, one for $2000, one for $1500, and one for $2500, then you average $2000). This becomes the current market price. You will need this number at the end of the game.

Chance Card

When a player lands on a chance card spot, he/she picks a chance card and follows the direction on the card. If they receive allowances or money, the EPA representative will give that player the proper amount required by the card.

Gain and Lose Pollution

Throughout the game, pollution output is continually changing. Players will gain and lose pollution as directed by the squares they land on.

Control Technologies

When you land on a control technology spot, you roll to determine which technology you purchase. Follow the chart below to see which roll equals which technology. These technologies will have a great price, but they will generate allowances for you later in the game. You may only purchase technologies when you land on these squares. You are not obligated to buy this technology. If you are continually landing on this spot, then you may continually purchase technologies and reap the benefit every time you pass start.

Roll	Technology
1	Wetscrubber
2	Baghouse
3	SCR
4	Reburning
5	Your Choice
6	Your Choice

The game is played for an allotted amount of time. At the end of the time, the players add up their money. Any pollution cards that do not have allowances are penalized by $5000 (the EPA price). For every allowance card that you have extra, you may add to your total the last market value of those allowances. For example, if during the last open market Bill sold Jack an allowance for $1500 then that is the current market of allowance cards so that if John finished the game with $10,000 cash and three extra allowances cards, his total worth would be $14,500. The winner is the individual with the highest worth.

THE ALLOWANCE GAME

POLLUTION
LOSE 1
CONTROL TECHNOLOGY
CHANCE ?
OPEN MARKET
GAIN 1 POLLUTION
CONTROL TECHNOLOGY
CHANCE ?
LOSE 1 POLLUTION
OPEN MARKET
GAIN 1 POLLUTION
CHANCE ?
CONTROL TECHNOLOGY
GAIN 1 POLLUTION
OPEN MARKET
LOSE 1 POLLUTION
CHANCE ?
CONTROL TECHNOLOGY
CHANCE ?
LOSE 1 POLLUTION
OPEN MARKET
START
CONTROL TECHNOLOGY
GAIN 1 POLLUTION
CHANCE ?

From *Clean Air Activities* by the Clean Air Conservancy. © 2003 by Humanics Learning

PLEASE MAKE 7 COPIES

★ **$500**	★ **$500**
★ **$500**	★ **$500**
★ **$500**	★ **$500**
★ **$500**	★ **$500**
★ **$500**	★ **$500**

PLEASE MAKE 7 COPIES

PLEASE MAKE 7 COPIES

$5000 · $5000 · $5000 · $5000 · $5000 · $5000 · $5000 · $5000 · $5000 · $5000

PLEASE MAKE 7 COPIES

ALLOWANCE

ALLOWANCE

ALLOWANCE

ALLOWANCE

ALLOWANCE

ALLOWANCE

ALLOWANCE

ALLOWANCE

ALLOWANCE

ALLOWANCE

ALLOWANCE

ALLOWANCE

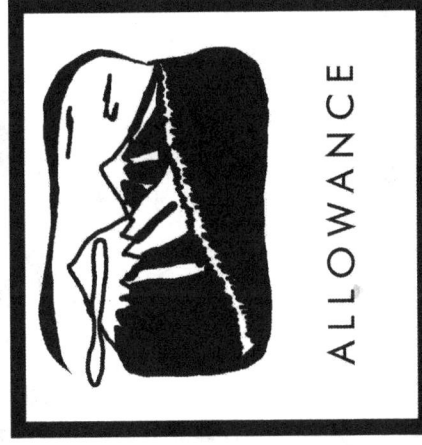

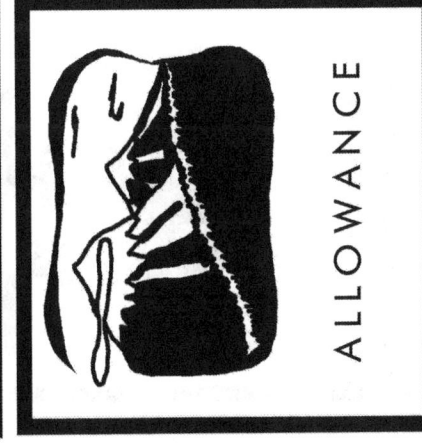

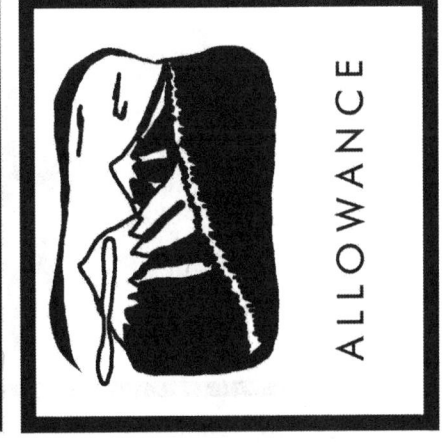

PLEASE MAKE 7 COPIES

POLLUTION

POLLUTION

POLLUTION

POLLUTION

POLLUTION

POLLUTION

POLLUTION

POLLUTION

POLLUTION

POLLUTION

POLLUTION

POLLUTION

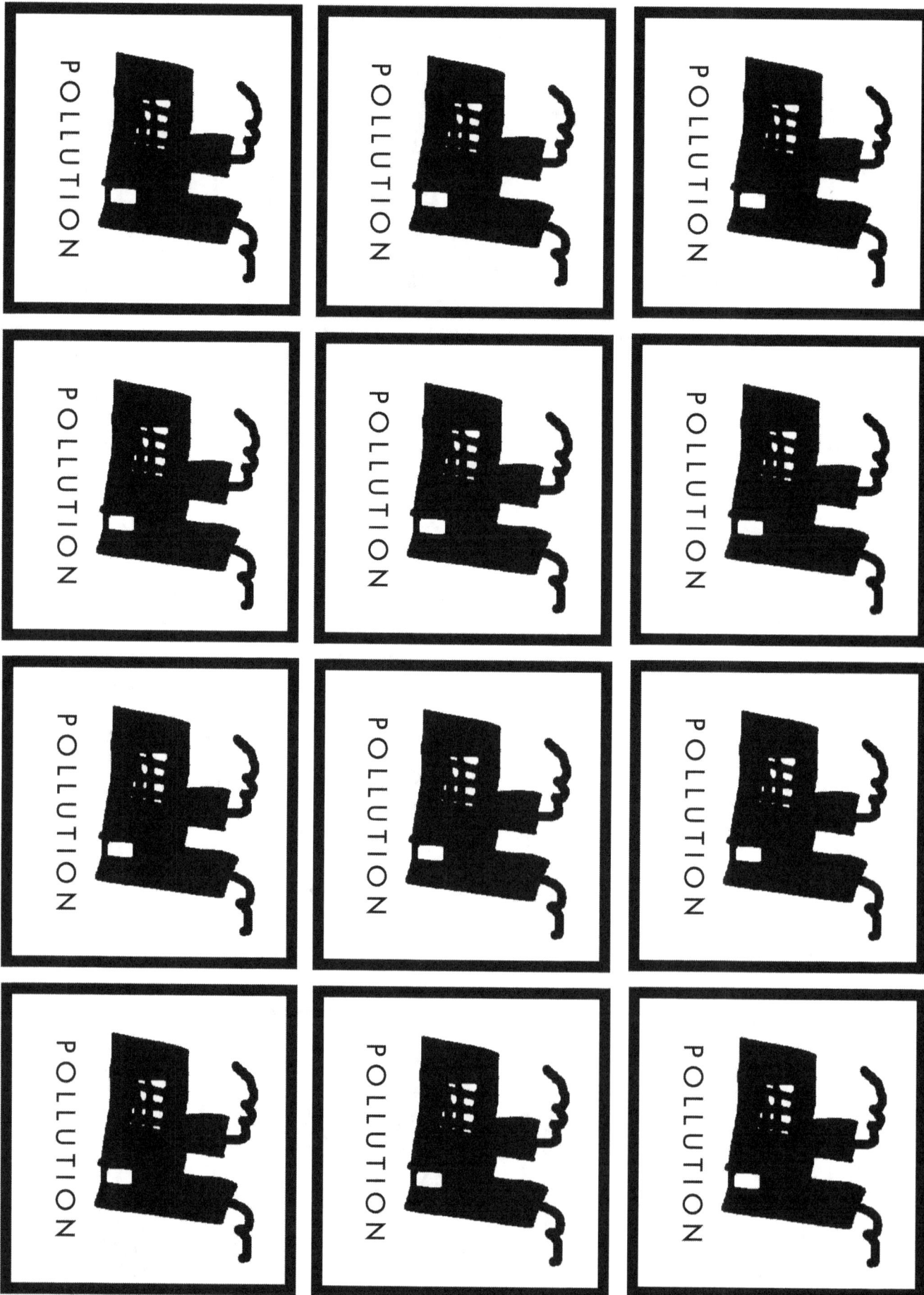

PLEASE MAKE 7 COPIES

CONTROL
TECHNOLOGY

CONTROL
TECHNOLOGY

CONTROL
TECHNOLOGY

CONTROL
TECHNOLOGY

CONTROL
TECHNOLOGY

CONTROL
TECHNOLOGY

CONTROL
TECHNOLOGY

CONTROL
TECHNOLOGY

CONTROL
TECHNOLOGY

CONTROL
TECHNOLOGY

CONTROL
TECHNOLOGY

CONTROL
TECHNOLOGY

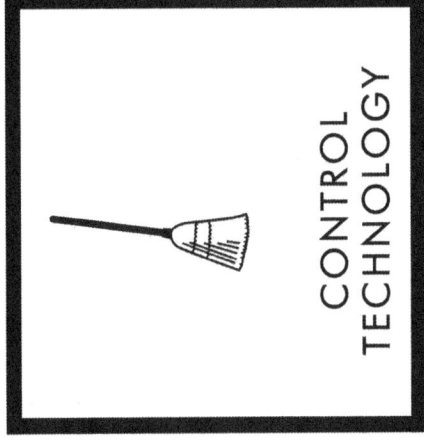

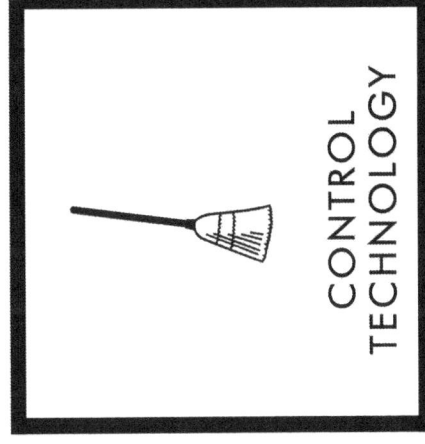

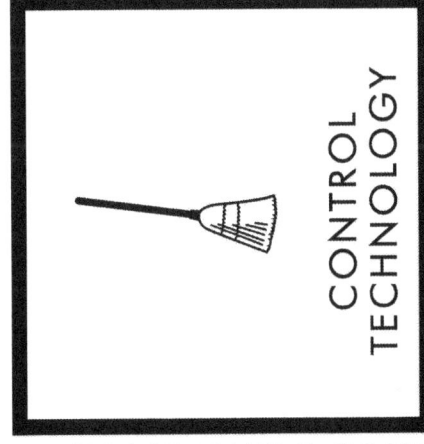

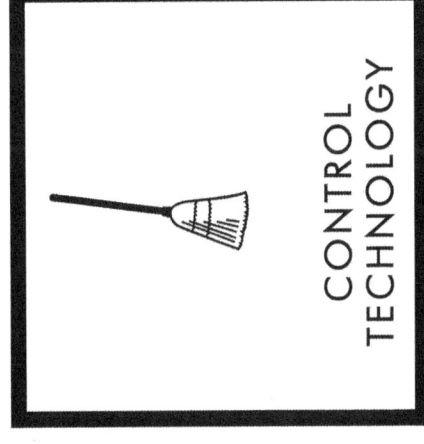

PLEASE MAKE 7 COPIES

CHANCE

CHANCE

CHANCE

CHANCE

CHANCE

CHANCE

CHANCE

CHANCE

CHANCE

CHANCE

CHANCE

CHANCE

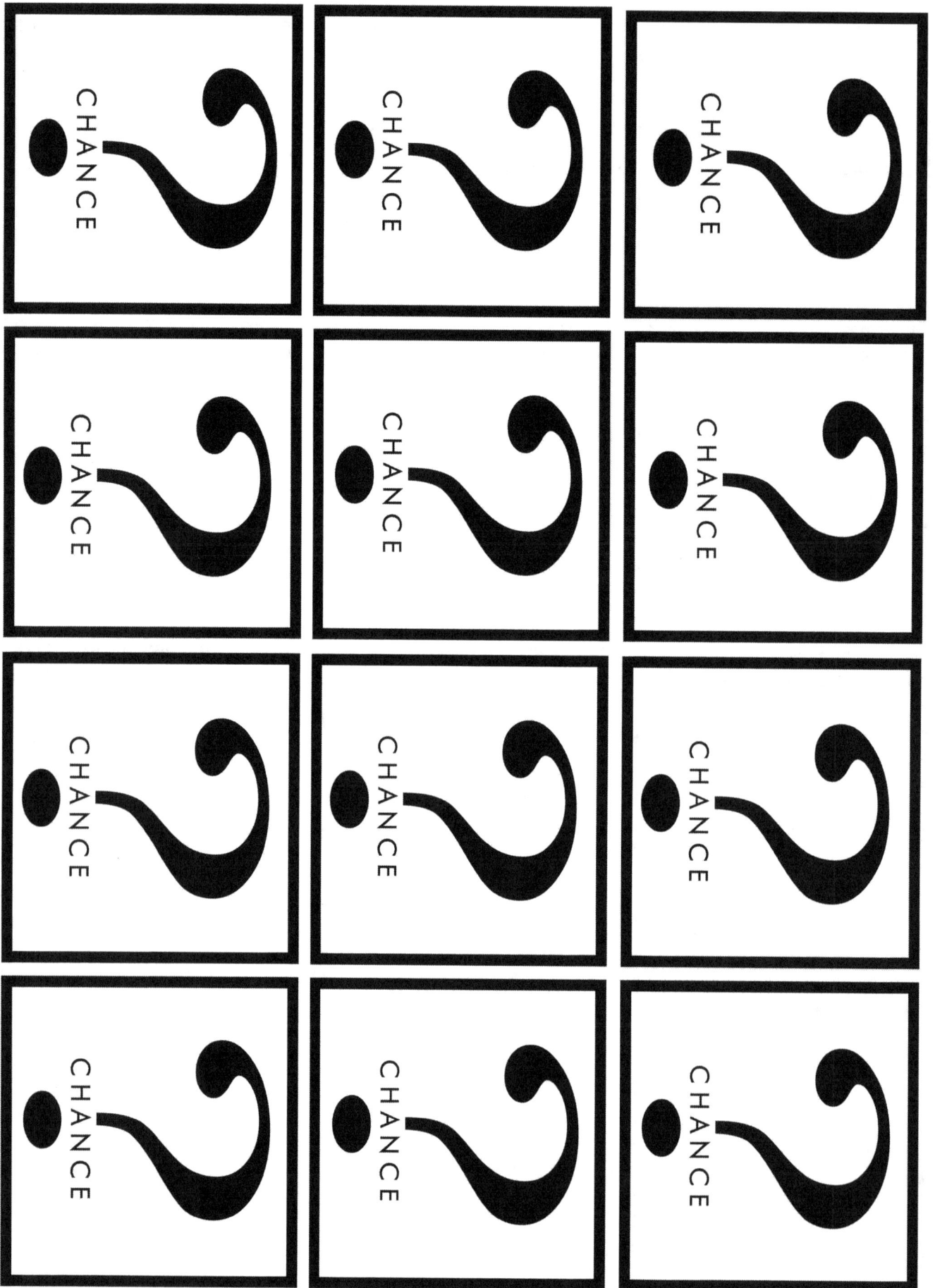

PLEASE MAKE 1 COPY BACK TO BACK WITH CHANCE SHEET

EPA audits your utility company, finds you pollute more than expected.

Fined twice market value.

($1000 if no market value)

Buy 1 allowance from player to your left for market price value.

($1000 if no market value)

Profits for this quarter worse than expected.

Lose $2000.

EPA audits your utility company, finds you in compliance.

Receive one free allowance card.

Go directly to start.

Collect $2000.

Buy a control technology if you want.

Profits for this quarter worse than expected.

Lose $5000.

Your customers lower energy consumption.

Lose one pollution card.

Very cold winter, strong demand for electricity.

Gain two pollution cards, but gain $2000 in revenue.

Profits for this quarter better than expected.

Receive $7500.

Clean Air Conservancy buys Allowances.

Lose one allowance card.

Gain Market Price

($1000 if no market value)

You can buy your choice of Control Technology.

Profits for this quarter better than expected.

Receive $5000.

PLEASE MAKE 1 COPY BACK TO BACK WITH CHANCE COVER

From *Clean Air Activities* by the Clean Air Conservancy. © 2003 by Humanics Learning

Renewables Project

STUDENT PROJECTS ON CLASS ENERGY USE

Background:
Within the curriculum module developed by The Clean Air Conservancy one of the major goals is to tie education about energy use and air pollution. To that end we recommend a number of issues be addressed including the science of air pollution, the health and environmental impacts of air pollution, the contribution individual's actions have on the problem, and some proactive things that students can do to specifically reduce pollution outcomes in their own lives.

Learning Objectives:
1. The link between daily individual behavior and pollution outcomes. Not all pollution is the fault of industry – what do individuals contribute?
2. The types of behavior that generate pollution and how much.
3. Connection between where you live and how you live impacting your energy use and pollution outcomes
4. Options available to individuals to change the amount of pollution they generate
5. Integrate these materials into a proposal to the school principal which outlines actions to be taken, costs, and the rate of payback of the expenditures, and the eventual pollution savings available.

Types of Activities:
1. Analyze your classroom for energy use and propose ways to make the classroom more energy efficient. What could be done for $2,500 and how quickly would the payback happen?
2. Examine the way all students are transported to school (vehicles used, type of trips etc). What are the pollution loads inherent and what kind of incentives could be developed to encourage more energy efficient transport?
3. Look at the way energy is used in the entire school. What types of activity can be taken to make the existing power plant more effective (insulating hot water pipes, insulation, better windows) and what would this save? Are there ways the school can employ alternative energy? Solar panels? Waste to energy? One building in Cincinnati has built highly energy efficient walls out of crushed soda cans and plastic wrap. Can that be used to face outer walls that are poorly insulated?

What the outcomes of Activity #1 should be:
1. Understand the difference in pollution outcomes from burning coal, oil and natural gas.
2. Understand the different areas where energy is lost from a building or room.
3. Understand the differences between energy saving technologies that could

be used to reduce energy use. It is also important that they understand the costs of the technologies and how the economic balance between cost and energy savings.

4. The student should be made aware of direct cost/benefits as well as the more indirect pollution costs – the damage from pollution is excluded because it isn't directly considered a cost.

Application of Activity #1

1. Conduct the classroom energy audit.
2. Calculate the pollution output from your classroom
3. Recalculate the audit based on a series of changes to the "box" – changes in windows, lights, insulation etc. – What would you have to do to become close to a zero discharge classroom? How many kilowatt hours would you save?
4. Calculate the costs of the changes you think are needed. How long would it take to pay back this investment?
5. Create a range of proposals – higher and lower cost fixes, the payback periods and the energy and pollution savings. Select a proposal and discuss it with the administration.

Possible Follow-ups to Reinforce the Lesson:

1. Discuss energy efficiency more broadly; as a school-wide project and as a home project.

Energy Efficiency Costs

The following list shows the source of the energy leakage, the range of alternatives, and a good estimate of the cost of each alternative. These estimates are based on the retail costs a homeowner would pay. Clearly, if the school district buys these items in bulk the result will be 25 to 40% less.

1. Windows – Windows are one of the largest sources of energy loss. Even though this is an expensive option, the savings is potentially great.

Double pane, gas filled windows	$15/sq ft of replacement window
High "e" glazed window	$20/sq ft of replacement window
Storm windows	$5/sq ft of replacement window

2. Insulation – Insulation is a very low cost option – typically for older buildings the insulation can be blown into the walls.

Cellulose blown into walls	$1/sq ft of insulation (adds r5)
Wall Board added to the walls	$3/sq ft of insulation (add r14)

3. Lighting – Lighting is a second high energy use area. The lighting is complicated by the need to replace both the lights and the fixtures.

From incandescent to compact florescent	$12/light

From florescent to compact florescent $12/light + $4/needed fixture.
Remember for every 4' of florescent tube replaced you have one compact florescent fixture)

4. Appliances – In most classrooms this is a minor issue except for computers. Estimate that at this point almost any new computer is energy star compliant. Given this, the cost is replacing old equipment with new ones.

Average of $1,200/machine.

The Awful 8: Introduction for the Play

PURPOSE:

To become aware of different air pollutants and their causes and effects.

OBJECTIVE:

The students will be able to list major air pollutants, what causes them, and their effects on people and the environment.

FOCUS:

After studying air pollution, students will present a play about the different pollutants.

MATERIALS:

Markers, yardsticks, large pieces of poster board, background information on air pollution, library books that cover air pollutants, materials for "costumes" copies of play for each student, video camera.

PROCEDURE:

Assign each part under the "Cast of Characters" and pass out copies of the play. Give the kids time to learn their lines, design costumes, and plan any special effects they might want to add.

After the group performs the play, review the eight major air pollutants by having each "pollutant" come out and take a bow. The Pollutants should state their name; what causes them; how they affect people, wildlife and the environment; and what people can do to help reduce this type of pollution. Or you can have the audience supply this information to see how much they learned from watching The Awful 8.

CLOSURE:

1. List ways we can prevent or reduce the types of air pollution mentioned in the play.
2. Brainstorm solutions to air pollution problems - be creative.

The Awful 8

A play about eight major air pollutants

CAST OF CHARACTERS:
(The number of characters and some suggestions for props and costumes are in parentheses.)

Connie Lung, reporter (1; props: microphone, notebook)

Harry Wheezer, reporter (1; props: microphone, notebook)

The Particulates (3; prop: dirt; costume: dirty jeans and brown t-shirts, smear dirt on face)

Carbon Monoxide (1; costume: sneakers, hat, trenchcoat, and sunglasses)

The Toxins (5; props: gasoline cans made from cardboard; costume: black clothing)

Sulfur Dioxide (1; prop: water gun or spray bottle filled with water; costume: torn t-shirt, yellow and white streamers attached to clothing)

Nitrogen Oxides (Nitros) (5; prop: dead branches; costume: each Nitro can wear one of the letters in "nitro")

Bad Ozone (1; costume: sunglasses, sophisticated clothing for a "big city look")

Good Ozone (1; costume: sunglasses and light-colored clothing with bits of cotton attached to represent clouds)

Chlorofluorocarbons (CFCs) (4; costume: heavy coats and jackets with the initials "CFC" stapled to the lapel and on the back, gloves and scarves)

EPA Scientists (2; prop: notebooks)

Carbon Dioxide (2; costume: t-shirts and shorts, black costume makeup wiped on clothing, legs and faces)

TIPS FOR PUTTING ON THE PLAY
• Have the Pollutants make picket signs by taping large pieces of poster board to yardsticks and writing slogans on the poster board. (See slogan suggestions in description of the play's setting.)
• If some kids prefer non-speaking roles, you can let them carry picket signs or

be camera people filming the report. They could also take on the responsibilities of stage manager, costume designer or set designer.
- Go over these pronunciations with the kids playing The Toxins; benzene (BEN-zeen), xylene (ZI-leen), toluene (TOL-you-een).
- If your audience is small, have Harry and Connie come up with some ways that people can help reduce air pollution at the end of the play.

SETTING

In front of the Environmental Protection Agency (EPA) building. The air pollutants are picketing the EPA. Some carry picket signs with phrases such as "Dirty Air! Let's Keep It That Way," "Down with the Clean Air Act" and so on. TV reporters Connie Lung and Harry Wheezer are at center stage. In turn, each pollutant comes over to be interviewed, while the other pollutants continue to picket in the background.

Connie: Hi! I'm Connie Lung.

Harry: And I'm Harry Wheezer. We're here at the Environmental Protection Agency to cover a late-breaking story. Eight of the world's worst air pollutants are picketing the EPA to protest clean-air legislation.

Connie: In tonight's special report, we'll give you the scoop on where these pollutants come from and the ways they can hurt people and other living things.

Harry: Our first interview is with the Particulates. (Particulates walk over, carrying signs and chanting.)

Particulates: Dust, soot and grime.
Pollution's not a crime
Soot, grime and dust,
The EPA's unjust!

Connie: (coughs) So - you're the Particulates.

Particulates 1(Soot): Yeah - I'm Soot, this is Grime and this is Dust.

Harry: You guys are those tiny bits of pollution that make the air look really dirty?
Grime: Yeah! Some of us are stirred up during construction, mining and farming. (throws some dirt in air)

Soot: But most of us get into the air when stuff is burned-- like gasoline in cars and trucks or coal in a power plant and even wood in a wood-burning stove!

Dust: And we just love to get into your eyes and make them itch and make your throat hurt and...

Grime: (interrupts) Come on, Dust, quit bragging! We gotta get back to the picket line. (Particulates return to picket line. Carbon Monoxide sneaks up behind Harry.)

Harry: Let's introduce the folks at home to our next pollutant, Carbon Monoxide. Hey, where did he go? Oh, there you are! Pretty sneaky, Carbon Monoxide!

Carbon Monoxide: Yeah, sneaking up on people is what I do best. I get into the air when cars and trucks burn fuel inefficiently - but you can't see or smell me.

Connie: Then how can we tell when you're around?

Carbon Monoxide: You'll find out when you breathe me in! I can give you a bad headache and make you really tired. (gives an evil laugh)

Harry: (yawns) Oh-- I see what you mean. Thanks for talking with us Monoxide. (yawns again)

(Carbon Monoxide returns to picket line.)

Connie: (checking notes) Next we'd like you to meet some of the most dangerous air pollutants-- The Toxins. (Toxins walk over, carrying signs and chanting.)

Harry: You Toxins are made up of all kinds of poisons. How do you get into the air?

Toxins 1: Hey, man, we come from just about everywhere. Chemical plants, dry cleaners, oil refineries, hazardous waste sites, paint factories...

Toxins 2: Yeah, and cars and trucks dump a lot of us into the air too. You probably don't know it, but gasoline is loaded with us toxins.

Toxins 3: Wow, that's for sure. There's benzene, toluene- all kinds of great stuff in gas.

Connie: Scientists say you cause cancer and other kinds of diseases. What do you think of that?

Toxins 4: They can't prove a thing!

Toxins 5: That's why we're here- to make sure you people don't pass any more laws that might keep us out of the air. C'mon, Toxins- we're outta here!

(Toxins return to picket line. Sulphur Dioxide walks over.)

Connie: Next we'd like you to meet Sulphur Dioxide. (Turns to face Sulphur Dioxide) I understand you just blew in from the Midwest.

Sulphur: Hey, I wouldn't miss this for all the pollution in New York City!

Harry: I'm sure the folks at home would like to know how you get into our air.

Sulphur: Well, heck, don't they read the newspapers? I've been making the front page at least once a week! Most of the time, I shoot out of smokestacks when power plants burn coal to make electricity.

Connie: And what kinds of nasty things do you do?

Sulphur: Nasty - that's me! (snickers) I think it's cool to make it hard for some people to breathe. And I can make trees and other plants grow more slowly. But here's the most rotten thing I do: When I get way up into the air, I react with oxygen in water in the sky, and presto! You get acid rain! (sprays water at the audience)

Harry: Acid rain is a big problem. It can hurt or kill fish and other animals that live in lakes and rivers and some scientists think it makes trees sick. Acid rain can even eat away at statues and buildings.

Sulphur: (proudly) That's right. Hey, I can even travel a long way to do my dirty work. If I get pumped out of a smokestack in Ohio, I can ride the wind for hundreds of miles and turn up as acid rain in Vermont!

Connie: I sure hope we can get rid of you soon, Sulphur Dioxide!

Sulphur: Good luck, guys! I gotta do some more picketing before I catch the next east wind!

(Sulphur Dioxide returns to picket line. Nitros walk over.)

Harry: (to the audience) He's really rotten!

Nitros: (all together) You think Sulphur Dioxide is rotten? You haven't met us!

Connie: You must be the Nitrogen Oxides.

Nitro 1: Just call us the Nitros for short. (turns to audience) Give me an "N!"

Audience and other Nitros respond: "N!"

From *Clean Air Activities* by the Clean Air Conservancy. © 2003 by Humanics Learning

Nitro 2: Give me an "I!"

Audience and other Nitros respond: "I!"

Nitro 3: Give me an "T!"

Audience and other Nitros respond: "T!"

Nitro 4: Give me a "R!"

Audience and other Nitros respond: "R!"

Nitro 5: Give me an "O!"

Audience and other Nitros respond: "O!"

Nitro 1: What's that spell?

Audience and other Nitros: NITRO!

Nitro 2: What's that mean?

Other Nitros: DIRTY AIR!

Harry: Hey, I didn't know pollutants could spell.

Nitro 4: Very funny, Harry.

Connie: So, how do you Nitros get into the air?

Nitro 5: We get airborne when cars, planes, trucks and power plants burn fuel.

Harry: And what happens once you're in the air?

Nitro 1: We can make people's lungs hurt when they breathe-- especially people who already have asthma.

Nitro 2: And, like Sulphur Dioxide, we react with water in the air and form acid rain.

Nitro 3: But we also make another form of pollution. And here she is - BAD OZONE! (Bad Ozone waves and walks over. Nitros return to picket line.)

Bad Ozone: Well, my friends the Nitros pour into the air and get together

with some other pollutants. As the sun shines on all these lovely pollutants, it heats them up - and creates me, Bad Ozone. And where there's ozone, there's smog.

Harry: (to audience) Smog contains a lot of ozone.

Connie: That's right, Harry. And smog can really make city life miserable. It can make your eyes burn, your head ache and it can damage your lungs.

Harry: But what I want to know is, if ozone is so bad, why are people worried about holes in the ozone layer? (Good Ozone walks in from offstage.)

Good Ozone: That low-level ozone is my rotten twin sister - she's just a good gas turned bad! I'm the good ozone that forms a layer high above the Earth. I help absorb the harmful rays of the sun.

Bad Ozone: (nastily to Good Ozone) So what are you doing here, sis?

Good Ozone: I'm here to support the clean air laws. If certain chemicals keep getting pumped into the atmosphere, I'll disappear. And without me, the harmful rays of the sun will kill some kinds of plants and give many more people skin cancer and eye disease!

Harry: But what kinds of chemicals are making you disappear?

Good Ozone: It's those terrible CFCs! (CFCs walk over from picket line.)

CFC 1: Hey, we're not so bad! People have used us CFCs in coolants for refrigerators and air conditioners for your home and car.

CFC 2: So what if we destroy a little bit of ozone? There's enough to last for years!

CFC 3: Yeah- who needs ozone anyway?

Good Ozone: People do! Tell them what else you CFCs are doing!

CFC 4: What's Ozone complaining about now - global warming? (EPA scientists walk in from offstage. Good and Bad Ozone walk offstage.)

Scientist 1: Excuse me, but did I just hear someone mention global warming?

CFC 2: Yeah. What do you want?

Scientist 2: We just happen to be experts on global climate change.

Connie: Are CFCs really changing the world's climate?

Scientist 1: Well, we're not positive. But over the past 100 years or so, people have been pouring gases, such as CFCs and carbon dioxide, into the air.

Scientist 2: And as they build up in the atmosphere, these gases may be acting like the glass in a greenhouse.

Scientist 1: That's right. They let the radiation from the sun in -- but they keep the heat from getting out. And this may be causing the Earth's climate to become warmer.

Harry: I've read that if the temperature goes up, sea levels may rise. Wow, some cities on the coast might be flooded some day!

Scientist 1: Well, nice talking with you all, but we've got to do some more research so that we can really nail these pollutants. (Points to CFCs. CFCs give scientists a dirty look, stick out tongues. Scientists walk offstage.)

CFC 1: Hey, we're not even the biggest cause of global climate change. You gotta talk to another of the big pollutants about that.

Harry: (checks notes) There's only one other pollutant on the list: Carbon Dioxide. (CFCs return to picket line. Carbon Dioxide 1 and 2 walk over.)

Dioxide 1: Did we hear you mention our name? We aren't really a bad gas, in the right amount. About a hundred years ago, there was just the right amount of us in the air.

Dioxide 2: But then people started burning more and more things - they built power plants that burn coal, and cars and trucks that burn gasoline. And they started cutting down and burning forests! Every bit of that burning releases extra amounts of us into the air.

Dioxide 1: As more and more of us got into the air, people started saying that the Earth was warming up-- because of us!

Dioxide 2: Yeah - like it's our fault! (to audience) The reason you're in such a mess is because you use so much fuel and cut down so many trees!

Connie: You're right, Carbon Dioxide. Maybe we should be doing a special report on people-- we're the ones who are really causing most air pollution.

Harry: But people can change! (turns to audience) How about you? Can you think of some ways that people can help fight air pollution? (Audience

responds with ideas, such as driving cars less, using less electricity, conserving forests, planting trees and so on.)

Connie: And that's the end of our special report. The bottom line? These air pollutants are a pretty tough bunch - but people help create much of them, and people can reduce the amounts that are in our atmosphere. Thank you and good night.

Pollutant curtain call.

The End.

From *Clean Air Activities* by the Clean Air Conservancy. © 2003 by Humanics Learning

Environmental Research Project

This project is a cooperative group research project. You will not be working alone. You and the other members in your group must ALL contribute to the project. At the end of the project, you will get a chance to evaluate the other members of your group, and they will evaluate your contribution. Your group and your topic will be assigned to you.

There are several parts to this project. Read through them carefully. Ask if you do not understand. It is important that you include every part in order to receive full credit. Here are the different parts that you MUST include:

I. The Report
A. You must include written information about the topic that you are studying, the history of the topic, and the problems it causes for the environment. Include as much information as you can. Refer to the list of key words on your topic sheet to help you get ideas of what to look for and what information to include. Write this information in paragraph form.
B. Include in your report a list of 10 to 20 relevant statistics related to your topic.
C. Include in your report a list of 15 to 25 **"Ways to Save..."** ideas to share with the class, your family and the community. These are things that we can do to lessen the pollution in our air. Your list should relate to your topic.
D. You will conduct an investigation that includes a science experiment. You will need to write your experiment using the Scientific Method. Include the steps of your project and the results in your report.
E. You need to **list all of the sources** that you used in your research. You must have between 10 to 20 sources. You should have some books, articles, and internet sites. Keep track of the title, author(s), date of publication, city of publication, and page numbers the information was taken from using this list.

II. The Poster
A. You must make a **quality poster** that looks professional. Your poster is separate from your report. Your can think of a message or slogan that relates to your topic. Your poster should be colorful and make people aware of the problem.
B. **Every group member** should help think of poster ideas.

III. The Presentation
A. You need to come up with a **four to five minute presentation** about your topic. Every member of your group should have a part in the presentation. You will state the topic or problem that you researched and tell

how it affects the environment and what we should do to correct it.

IV. Other Parts
A. Some topics will require that you do **other things**. You will work with each group individually to come up with some other things to add to their projects.
B. You may find that you want to include something else in your project. Include any additional information that you find that is revelant to your topic.

Environmental Research Project

Topic Sheet and Checklist

YOUR TOPIC IS: POLLUTION AND WILDLIFE

List of Key Words: (List some of your own.)

plants	habitat	deforestation
pesticides	animals	vegetation
web of life	endangered species	ecosystem
preservation	energy from the sun	tropical rainforest
		biointensive farming
_____	_____	
_____	_____	_____

Checklist:

_____ We have included written information about our topic, its history, and the problems it causes for the environment.

_____ We have included 10 to 20 relevant statistics.

_____ We have included 15 to 25 "Ways to Save . . ."

_____ We conducted an investigative science experiment. We used the scientific method. We included the steps and results of our experiment.

_____ We have been keeping track of the titles, authors, dates of publication, city of publication, and page numbers of all sources.

_____ We created an awesome, quality poster that is colorful.

_____ We even did extra work to make our research project stand out!

From *Clean Air Activities* by the Clean Air Conservancy. © 2003 by Humanics Learning

Environmental Research Project

Topic Sheet and Checklist

YOUR TOPIC IS: ACID RAIN

List of Key Words: (List some of your own.)

corrosion	sulfuric acid	pH levels
ground water	the Water Cycle	deterioration
damage to trees	damage to animals	nitrogen oxide
nitric acid	movement of pollution	sulfur dioxide
		mercury

_____   _____   _____

_____   _____   _____

Checklist:

_____ We have included written information about our topic, its history, and the problems it causes for the environment.

_____ We have included 10 to 20 relevant statistics.

_____ We have included 15 to 25 "Ways to Save . . ."

_____ We conducted an investigative science experiment. We used the scientific method. We included the steps and results of our experiment.

_____ We have been keeping track of the titles, authors, dates of publication, city of publication, and page numbers of all sources.

_____ We created an awesome, quality poster that is colorful.

_____ We even did extra work to make our research project stand out!

Environmental Research Project

Topic Sheet and Checklist

YOUR TOPIC IS: AUTOMOBILES AND THE ENVIRONMENT

List of Key Words: (List some of your own.)

transportation	emissions control	gas mileage
ride sharing	auto industry	sulfur dioxide
exhaust	smog	carbon dioxide
industry	catalytic converter	lead
_____	_____	nitrogen oxide
_____	_____	_____

Checklist:

_____ We have included written information about our topic, its history, and the problems it causes for the environment.

_____ We have included 10 to 20 relevant statistics.

_____ We have included 15 to 25 "Ways to Save . . ."

_____ We conducted an investigative science experiment. We used the scientific method. We included the steps and the results of our experiment.

_____ We have been keeping track of the titles, authors, dates of publication, city of publication, and page numbers of all sources.

_____ We created an awesome, quality poster that is colorful.

_____ We even did extra work to make our research project stand out!

From *Clean Air Activities* by the Clean Air Conservancy. © 2003 by Humanics Learning

Environmental Research Project

Topic Sheet and Checklist

YOUR TOPIC IS:
ENVIRONMENTAL ACTION GROUPS AND LEGISLATION

List of Key Words: (List some of your own.)

The Clean Air Conservancy	green markets	lobby/lobbyist
EPA	The Sierra Club	Greenpeace
legislation (laws)	environmentalist	activist
ecology	Clean Air Act of 1990	conservationist
		Earth Day
_____	_____	
_____	_____	_____

Checklist:

_____ We have included written information about our topic, its history, and the problems it causes for the environment.

_____ We have included 10 to 20 relevant statistics.

_____ We have included 15 to 25 "Ways to Save . . ."

_____ We conducted an investigative science experiment. We used the scientific method. We included the steps and results of our experiment.

_____ We have been keeping track of the titles, authors, dates of publication, city of publication, and page numbers of all sources.

_____ We created an awesome, quality poster that is colorful.

_____ We even did extra work to make our research project stand out!

Environmental Research Project

Topic Sheet and Checklist

Your Topic is: POLLUTION AND ENERGY

List of Key Words: (List some of your own.)

conservation	carbon dioxide	florescent bulbs
solar power	wind power	natural gas
nuclear power	fossil fuels	geothermal energy
electricity	power plants/utilities	Industrial Revolution
_____	fuel cells	recycling
_____	_____	_____

Checklist:

_____ We have included written information about our topic, its history, and the problems it causes for the environment.

_____ We have included 10 to 20 relevant statistics.

_____ We have included 15 to 25 "Ways to Save . . ."

_____ We conducted an investigative science experiment. We used he scientific method. We included the steps and results of our experiment.

_____ We have been keeping track of the titles, authors, dates of publication, city of publication, and page numbers of all sources.

_____ We created an awesome, quality poster that is colorful.

_____ We even did extra work to make our research project stand out!

Environmental Research Project

Topic Sheet and Checklist

YOUR TOPIC IS: OZONE AND HEALTH

List of Key Words: (List some of your own.)

ground level ozone	bronchitis	National Ambient Air
Ozone Action Days	stratosphere	Quality Standard
emphysema	summer	Volatile Organic
Chlorofluorocarbons (CFC's)	asthma	Compounds (VOCs)
_____	_____	_____
	_____	_____

Checklist:

_____ We have included written information about our topic, its history, and the problems it causes for the environment.

_____ We have included 10 to 20 relevant statistics.

_____ We have included 15 to 25 "Ways to Save . . ."

_____ We conducted an investigative science experiment. We used the Scientific Method. We included the steps and results of our experiment.

_____ We have been keeping track of the titles, authors, dates of publication, city of publication, and page numbers of all sources.

_____ We created an awesome, quality poster that is colorful.

_____ We even did extra work to make our research project stand out!

From *Clean Air Activities* by the Clean Air Conservancy. © 2003 by Humanics Learning

Environmental Research Project

Topic Sheet and Checklist

YOUR TOPIC IS: AIR POLLUTION AND VOLATILE ORGANIC COMPOUNDS

List of Key Words: (List some of your own.)

sulfur dioxide	particulate matter	allowances
wet scrubber	mixing of compounds	nitrogen oxides
primary locations	causes of air pollution	Volatile Organic
industry	Clean Air Act of 1990	Compounds (VOCs)
legislation (laws)	_____	_____
	_____	
_____		

Checklist:

_____ We have included written information about our topic, its history, and the problems it causes for the environment.

_____ We have included 10 to 20 relevant statistics.

_____ We have included 15 to 25 "Ways to Save . . ."

_____ We conducted an investigative science experiment. We used the Scientific Method. We included the steps and results of our experiment.

_____ We have been keeping track of the titles, authors, dates of publication, city of publication, and page numbers of all sources.

_____ We created an awesome, quality poster that is colorful.

_____ We even did extra work to make our research project stand out!

Environmental Research Project

Topic Sheet and Checklist

YOUR TOPIC IS: CLIMATE AND WEATHER

List of Key Words: (List some of your own.)

global warming	climate	wind speed
wind direction	the green house effect	stratosphere
ozone	atmosphere	skin cancer
thermal inversion	UV rays	weather
_____	_____	_____
_____	_____	_____

Checklist:

_____ We have included written information about our topic, its history, and the problems it causes for the environment.

_____ We have included 10 to 20 relevant statistics.

_____ We have included 15 to 25 "Ways to Save . . ."

_____ We conducted an investigative science experiment. We used the Scientific Method. We included the steps and results of our experiment.

_____ We have been keeping track of the titles, authors, dates of publication, city of publication, and page numbers of all sources.

_____ We created an awesome, quality poster that is colorful.

_____ We even did extra work to make our research project stand out!

Topic One: Pollution in the Air
ANSWER KEY

1. **N**ame four sources of Air Pollution.

a) Cars

b) Utilities

c) Industry

d) Small utility engines (lawn mowers, weed eaters)

2. **N**ame four pollutants created by electricity generation.

a) Nitrogen Oxide (NO_x)

b) Sulfur Dioxide (SO_2)

c) Carbon Dioxide (CO_2)

d) Particulate Matter

3. **W**hat is particulate matter? **N**ame examples of particulate matter.
Particulate Matter is small dust less than 10 microns in size emitted by burning matter. Size is important because the smaller the particle the more easily it evades the defenses of the lungs. Examples include soot, fly ash, metal flake, sulfate particles, and dust.

Define the following vocabulary words. Write on back if necessary.

1. carbon dioxide (CO_2) – Carbon Dioxide is an odorless, colorless gas that is a naturally occurring part of the environment. It is emitted when coal is burned to generate electricity. It also causes global warming.

2. nitrogen oxide (NO_x) – Harmless compounds created through the burning of coal. When combined with VOC's and sunlight it forms Ozone.

3. sulfur dioxide (SO_2) – A byproduct of coal burning; one of the major contributors to the formation of acid rain.

4. volatile organic compounds (VOC's) – carbon based compounds that readily evaporate into the air. These compounds tend to be carcinogenic and serve as Ozone precursors. Unburned gasoline, benzene, and xylene are examples.

5. particulate matter – Tiny particles in the atmosphere that can be solid or liquid. PM includes dust, dirt, soot, smoke, and tiny particles of pollutants.

Topic Two—Pollution Effects in the Environment
ANSWER KEY

1. **W**hat is acid rain? **W**here does it come from? **H**ow does it form?
   Acid rain is precipitation with high levels of SO_2. The combination of SO_2 and water leads to the production of sulfuric acid. The SO_2 is almost exclusively formed from the emission of coal burning power plants.

2. **W**hat is one of the major contributor to acid rain formation?
   Acid rain is formed by SO_2, H_2O and heat. Coal burning utilities are also a major contributor.

3. **W**hat are some of the environmental impacts of acid rain?
   Impacts include plant damage (roots are burned), heavy metals leaching out of the soil. Lowered stream pH inhibits aquatic reproduction.

4. **W**hat is the difference between tropospheric (ground level ozone) and stratospheric ozone?
   Tropospheric ozone is formed on the ground and is difficult to breathe. Stratospheric ozone is found in the atmosphere and protects life.

5. **W**hat causes tropospheric ozone?
   Nitrogen Oxide mixed with Volatile Organic Compounds (VOC's) on hot, windless days creates ground level ozone.

6. **N**ame three ways you can help limit the amount of ozone on Ozone Action Days.
   Buy gasoline after seven pm, do not cut the lawn until evening, walk, and turn down air conditioners.

7. **W**hat is an Ozone Action Day?
   Known as a smog alert in some parts of the country, an Ozone Action Day is a warning to modify energy use to prevent ozone formation. It is also a warning to limit physical activity in the middle of the day because ozone is a powerful lung irritant.

8. What is Global Climatic Change? What causes it? How can it be prevented?
   Global Climatic Change is the increasing level of energy trapped in the atmosphere. The energy is trapped by the increasing concentration of greenhouse gases in the atmosphere. These gases reflect energy that would normally radiate back into space, back to the planet. The result is increasing energy levels. It can be reversed by reducing the amount of greenhouse gases released into the environment.

9. What is the Greenhouse effect?
   It is the reflected energy in the atmosphere that increases the severity of normal climatic patterns. This can lead to more powerful storms, temperature shifts, and shifts in long-term climatic patterns (drought).

10. Describe what happens during temperature inversion. Why is this harmful to the environment?
    In a temperature inversion, a bubble of warm air covers a region, redirects winds and traps pollutants in a bubble close to the earth's surface. There have been fatalities from inversion when toxic pollutants have not been able to escape to the higher atmosphere.

Define the following vocabulary words.
1. global warming – A term used to describe the effects of too much carbon dioxide in the atmosphere. It is the reality of higher energy levels in modern weather phenomena.

2. smog – The result of the combination of pollutants and atmospheric humidity which is also known as ozone. It also diminishes visibility.

3. asthma – A lung disease which makes breathing difficult. This is exaggerated by ozone and other pollutants in the atmosphere.

4. Ozone Action Days – Alerts issued by local air agencies to warn individuals about dangerous levels of ground level ozone. In some communities they are known as "smog alerts."

Topic Three: Regulating Pollution
ANSWER KEY

1. **W**hat are the Clean Air Act Amendment of 1990?
   This is legislation which established a sulfur market to control pollution emissions.

2. **W**ho is the Clean Air Conservancy?
   The CAC is a nonprofit environmental organization dedicated to finding market solutions for pollution control.

3. **W**hat is a pollution allowance?
   This gives a utility company permission to emit one ton of sulfur dioxide pollution into the air.

4. **N**ame two ways that you can help control air pollution:
   Buy pollution allowances, reduce energy use, walk, ride a bike, and find alternative means for generating electricity.

5. **W**hat is the EPA?
   The Environmental Protection Agency is charged with implementing legislation targeted at protecting the environment. It is also the responsibility of the EPA to monitor environmental conditions and make recommendations for improvement to Congress and the President.

Define the following vocabulary words. Write on back if necessary.

1. Clean Air Act – Legislation passed by congress intending to limit the amount of pollution utilities emitted. This is difficult to enforce.

2. Clean Air Act Amendments – Changes made to the earlier law, which helped tighten pollution regulation. It established the sulfur dioxide pollution market in the United States.

3. Environmental Protection Agency (EPA) – A government agency charged with monitoring the quality of the environment in America. EPA usually implements and enforces the regulations passed by Congress.

4. Emissions standards – The allowable amount of pollutants in the air, established by the EPA.

5. Pollution Allowances/permits – This gives utility companies permission to emit pollutants into the atmosphere for a certain price.

Topic Four—Energy Efficiency
ANSWER KEY

1. Name three ways you can limit the amount of electricity you use.
   Turn off lights, change the types of light bulbs used, buy more energy efficient products, and insulate your home.

2. What is the difference between incandescent, florescent and compact florescent light bulbs? Which uses the most electricity?

   Incandescent lighting is old style technology where a fin wire, called a filament, burns as an electric current passes through it.

   Florescent lights have an electric arc that passes through a set of gases which glow.

   Compact florescent lights is a more energy efficient version of florescent lights that use less energy.

   A typical incandescent bulb uses 100 watts/hour. A florescent tube consumes approximately 25 watts/hr. A compact florescent light bulb uses 17 watts/hr for comparable light.

3. Name four types of alternative sources of energy.
   Hydro, wind, passive solar, active solar, waste gas recovery, fuel cells, and geothermal energy.

4. What is recycling?
   Recycling is taking a product and using it again.

Define the following vocabulary words. Write on back if necessary.
1. hydropower – Using water to generate electricity.
2. fuel cells – Using hydrogen passed through a thin membrane to generate electricity.
3. nuclear energy – Splitting atoms to generate electricity.
4. recycling – The act of reusing a product or using it differently. This is important in conserving the resources we have.
5. natural gas – A fossil fuel that is burned to generate electricity. It is cleaner than burning coal.
6. solar energy – Using the energy of the sun to generate electricity.
7. wind power – Using the power of the wind to generate electricity.

Topic Five—Control Technology
ANSWER KEY

1. **W**hat is a control technology?

A control technology is a designated technology that meets EPA criteria for pollution reductions. The three levels are RACT (Reasonable Available Control Technology), BACT (Best Available Control Technology), and MACT (Maximum Available Control Technology).

2. **W**hy do industries need to use control technologies?

Control technology is defined by EPA as the best regulatory means of correcting a pollution problem. The three levels are defined by the cost; the greater the health hazard, the more rigorous the technology that is demanded to fix the problem.

3. **M**ark the technologies that are used to remove particulate matter with a "P," Nitrogen Oxide with an "N," and Sulfur Dioxide with an "S."

S   Wetscrubber
P   Baghouse
S   Dry Scrubber
N   Low-NOx Burner
N   Selective Catalytic Reduction (SCR)
P   Electrostatic Precipitator
N   Selective Non-Catalytic Reduction (SNCR)

4. **D**oes a control technology completely clean the pollution industry creates? Explain.

No control measure completely cleans the air. The best we are aware of is approximately 98% reductions.

5. **I**f industry is jut part of the problem, what can WE do to control the amount of air pollution we cause?

Reduce energy use, produce your own energy, purchase pollution allowances, change how and what you buy and your transportation.

Define the following vocabulary words.

1. Selective Catalytic Reduction (SCR) – This uses a chemical reaction to break down the NO_x present in the flue gas released after combustion.
2. Low-NO_x combustion – A combustion control which lowers the tempeature used to burn coal so that NO_x cannot form.
3. Selective Non-Catalytic Reduction (SNCR) Control technology- Using science to control pollution emissions.
4. Overfire Air - Diverting a portion of the oxygen needed for complete combustion from the burners to an upper furnace.
5. Wet Scrubber – Wet flue gas desulfurization. It captures the sulfur emissions into a liquid compound before they can enter the atmosphere.
6. Electrostatic Precipitator – Used to control particulate matter, an EP uses an electrically charged plate to collect particulate matter.
7. Baghouse- Like a vacuum cleaner, it collects particulate matter before it can enter the atmosphere.

Carbon Dioxide and Air Pollution

PURPOSE:
To test for different concentrations of carbon dioxide gas.

OBJECTIVE:
1. Students will observe the effects of carbon dioxide gas.
2. Students will test for the presence of carbon dioxide gas.
3. Students will compare concentrations of carbon dioxide gas.
4. Students will conclude high concentrations of carbon dioxide gas are unhealthy for human beings.

MATERIALS:
matches
baking soda
vinegar
bromothymol blue (BTB)
straw
candle
clear cups, glasses, or beakers for each group of students
round balloons

BACKGROUND:
Carbon dioxide is produced when vinegar and baking soda are mixed. One can then "pour" the gas (carbon dioxide is heavier than air) over a flame and extinguish it. The gas will replace the oxygen and cause the flame to go out. BTB is an indicator for carbon dioxide. BTB will change from dark blue to light blue, or green to yellow, depending on the concentration of carbon dioxide. Students can test for the concentration of carbon dioxide using the gas produced when they exhale and when vinegar and baking soda are mixed.

PROCEDURE:
1. Demonstrate how carbon dioxide gas will extinguish a flame. Mix vinegar and baking soda in a beaker. Tilt the gas over a burning candle. (Do not tilt the beaker or bottle so much that the liquid runs out and extinguishes the flame.) Have students explain why the flame went out. Discuss each theory. Repeat the demonstration, explaining each step. (See explanation in background section.) Explain to students that they will create this same gas for an experiment.
2. Pour 50 ml of vinegar in a beaker or narrow-necked bottle. Place 45 grams of baking soda in a balloon.
3. Tip the balloon so the baking soda falls into the vinegar. Observe the inflation of the balloon. Twist the balloon shut so that the gas will not escape.
4. Put a straw into the balloon; a little carbon dioxide might escape.

5. Pour 50 ml of BTB solution into a clear container and place the straw with the balloon into the solution.
6. Allow the carbon dioxide to be released into the BTB solution.
7. Have students observe and infer what has happened. Have students record results in a science journal or data collection sheet.

Possible answers:
• BTB is a test for carbon dioxide
• Carbon dioxide was created when the baking soda and vinegar combined.
• A record of the color change.

1. Ask students to name the gas that animals exhale. Have students repeat the procedure to test for the concentration of carbon dioxide gas they exhale. Ask them to predict if the results will be the same or different. Will human exhaled air have a higher concentration of carbon dioxide gas than air that created when vinegar and baking soda are mixed? Observe and record results.

EVALUATION:

Some people think the levels of carbon dioxide in our atmosphere are too high. They are afraid Earth's temperature will rise making it unhealthy for life. The World Resources Institute says electric utility companies, industry, businesses, homes, and transportation cause carbon dioxide levels to build up in our atmosphere. Using pictures and words, explain how you could help reduce the levels of carbon dioxide.

Harmful Effects of Acid Rain

PURPOSE:
To demonstrate the harmful effects of an acidic solution. (Vinegar is a dilute solution of acetic acid.)

OBJECTIVE:
The student will become aware of the harmful effects of acid rain.

MATERIALS:
vinegar
water
2 medium sized eggshell pieces
2 small green leaves
two paper clips
two containers with lids

PROCEDURE:
1. Before activity, make predictions. If vinegar contains acid, then how will some items placed in vinegar change? If these items were placed in water, would they change in the same ways as in vinegar?
2. Pour vinegar in one container. Place an eggshell piece, a leaf, and a paper clip in the container. Put the lid on the container.
3. Pour water in the other container. Place an eggshell, a leaf, and a paper clip in this container. Put the lid on the container.
4. Let the two sealed containers sit overnight.
5. Remove the container lids. Observe any changes that took place in the two containers. Write down observations.

RESULTS:
In the container of water, the items will not show noticeable changes. In the container of vinegar, the eggshell will be soft, the leaf will have brown spots on it, and the paperclip will not show a noticeable change. This activity indicates that acidic solutions can be harmful.

EXTENSIONS:
Measure the acid in several solutions using inexpensive pHydrion papers (pH papers). Suggestions for solutions to be tested are:

lemon juice (pH of 2.0)
vinegar (2.2)
apple juice (3.0)
tomato juice (4.2)
milk (6.2)
pure water (7.0).

Compare the solution pH values with acid rain (below 5.6) and normal rain (above 5.6). Explain that some foods we eat have healthy acids like citric acid, which is not harmful. However, there are stronger acids which are the products of factories and industries that are harmful.

How To Make An Electrostatic Precipitator

MATERIALS, EQUIPMENT, AND PREPARATION
plastic tube (fluorescent light tube)
wire coat hanger
plastic grocery bag
electric blow dryer
punch holes, black pepper or rice krispies

The electrostatic precipitator works on the principle of a static electric charge attracting particles where they are removed.

A 2-foot plastic tube in which fluorescent lights are stored can be used to simulate an electrostatic precipitator. The plastic tube can be charged by running a coat hanger with a plastic grocery bag attached to it. (See diagram B)

As it moves through the tube the plastic bag strips the negatively charged electrons from the inside of the tube making the overall net charge positive. Anything that has a negative charge will be attracted to the tube because opposites attract.

Hold the tube over some punch holes, black pepper, or rice krispies. Hold an electric hair dryer so the air stream blows across the top of the tube. The air mass creates a low pressure area at the top and the greater air pressure at the bottom pushes the punch holes up the tube. (This is called Bernoulli's principle.

RESULTS:
If the tube is charged, the punch holes will stick to the sides.
This activity can be used to study static electricity.
If the tube is not charged, the holes will shoot out in a spray.
This activity can be used to study Bernoulli's principle.

EXTENSION:
Balloon Activity

MATERIALS:
pepper or ashes
balloons

PROCEDURE:
Give each student an inflated balloon and some black pepper. Rub the balloon on your hair or with a piece of cloth. Hold the balloon over the pepper on your desk.

What happens to the pepper?

Ask the students what produces air pollution. Discuss that industry is just one producer of air pollution. Ask what kinds of pollutants are produced by industry. Discuss that particles (called particulate matter) can be captured before they enter the atmosphere by an electrostatic precipitator. Demonstrate with the plastic tube and black pepper how particles are attracted to the sides of the tube much like the pepper was attracted to the balloon.

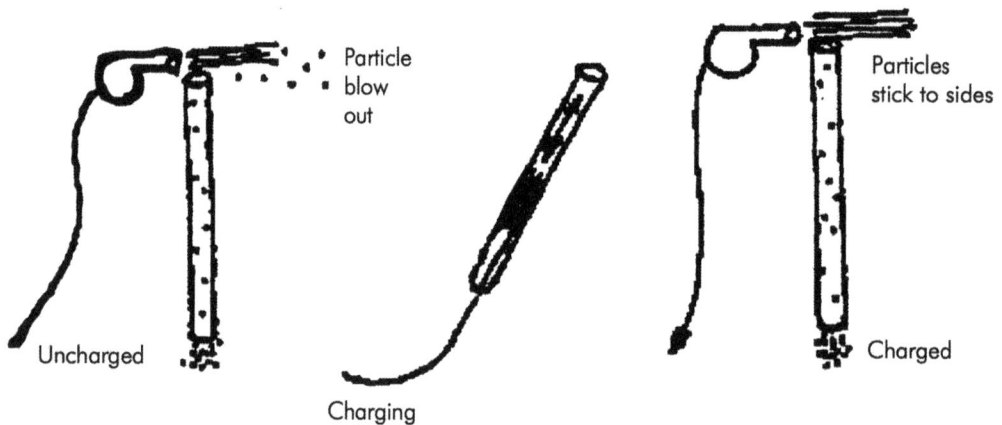

Particle
blow
out

Uncharged

Charging

Particles
stick to sides

Charged

FIGURE 2
Electrostatic Precipitator

How to Make a Wet Scrubber

PURPOSE:
To become familiar with a wet scrubber by building and using one.

MATERIALS:
paper towels (or other combustible paper)
12-cm piece of glass tubing
three five cm pieces of glass tubing
three 500-ml flasks
two glass impingers (glass tubing drawn at one end to give it a smaller diameter so as to let out smaller bubbles.
heat source (burner or hot plate)
three two-hole rubber stoppers (of a size to fit the mouths of the flasks)
two 30-cm pieces of rubber tubing
ring-stand apparatus
vacuum source

BACKGROUND INFORMATION:
The wet scrubber is one of the most common pollution control devices used by industry. It operates on a very simple principle: a polluted gas stream is brought into contact with a liquid so that the pollutants can be absorbed.

PROCEDURE:
1. Set up the apparatus as shown in figure one. Put a paper towel in a 500-ml flask and place this above the burner.
2. Using a two-hole stopper that makes an airtight seal with the flask, insert a 12-cm section of glass tubing through one of the holes. The tubing should reach to approximately 1.2-cm from the bottom of the flask.
3. Insert a five cm piece of glass tubing into the other hole of the stopper.
4. Connect a 30-cm piece of rubber tubing to the five cm piece of glass tubing, making sure an air-tight seal exists.
5. Fill a second 500-ml flask approximately 3/4 full of water. Using a second double-hole stopper, put a 5-cm piece of glass tubing into one of the holes, and insert the glass impinger into the other.
6. Construct a third flask like the second.
7. Connect the rubber tubing and heat the first flask (combustion chamber) until smoke appears.
8. Put a vacuum on the third flask to draw a stream of smoke through the second flask (the wet scrubber). If smoke collects in the second flask above the water, a second scrubber can be added.
9. Ask the students if particles are the only pollutants produced by industry. Discuss how a wet scrubber collects not only particulate matter but also captures waste gases. Demonstrate how the water scrubber works. Tell your students that the white plume they see coming from a smokestack may really be steam coming from a water scrubber.

10. After observing the wet scrubber, answer the following questions:

- Why does the water in the wet-scrubber change color?
- Why does the wet-scrubber have an impinger (in other words, why is it important for small bubbles to be formed)?
- What does the scrubber filter out of the air? Not filter out?

Suggest ways to dispose of the pollutants that are now trapped in the water.

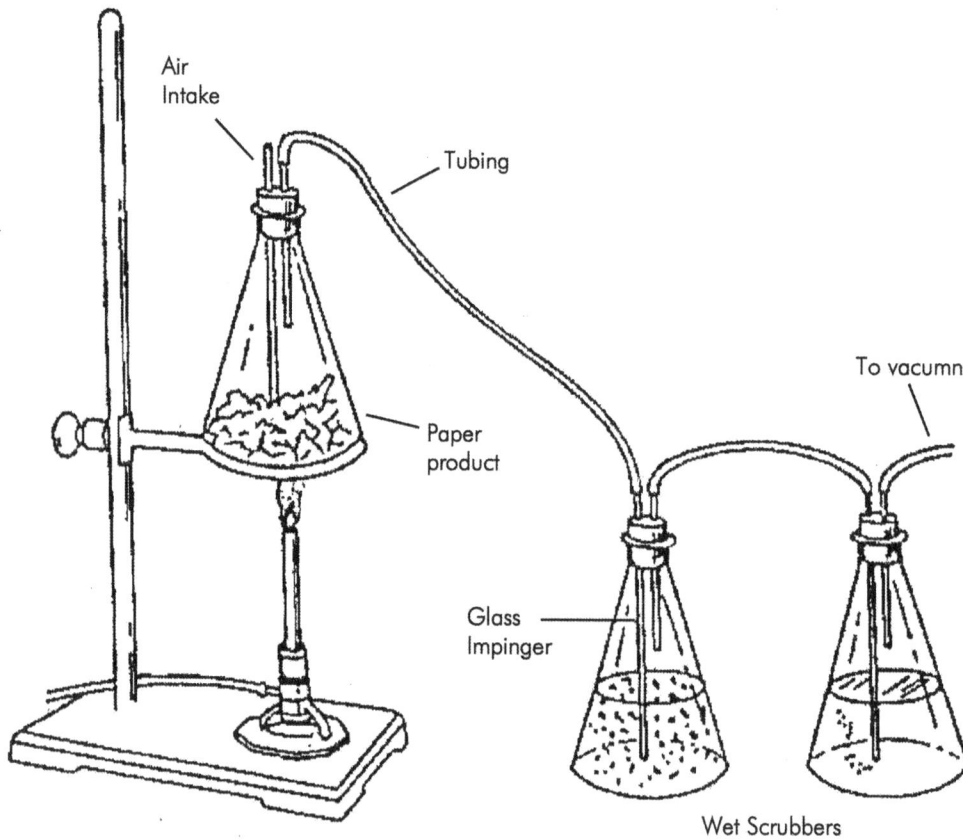

FIGURE 1
Wet Scrubber

The Rubber Band Air Test

PURPOSE:
To discover effects of air pollution.

Show students a variety of pictures showing air pollution. Explain that sometimes air pollution is very easy to see, but can also be very hard to detect.

MATERIALS:
Four small rubber bands, one wire clothes hanger, magnifying glass

BACKGROUND:
The atmosphere is almost completely made up of invisible gaseous substances. Most major air pollutants are also invisible, although large amounts of them concentrated in areas such as cities can be see as smog. Sometimes we can see smog, but most of the time it's invisible. Carbon dioxide is one of the gases that contributes to the greenhouse effect and is the primary gas that makes up smog. This lesson will help you see that air pollution is all around you even if you can't see it. It will also give you an idea of the effect air pollution has on our earth.

PROCEDURE:
1. Discuss causes and effects of air pollution.
2. Bend the hanger so that when you stretch the rubber bands over the hanger, they are tight.
3. Hang the hanger outdoors in a shady place so it's out of the sun and leave it there for two weeks.
4. When two weeks are up, look at the rubber bands. Do they look the way they did before, or are they cracked? Check with the magnifying glass too.
5. Touch the rubber bands. Do they feel the way they did before, or are they hard? If they look and feel the way they did before, then the air is quite clean. If they look cracked and feel hard, then the air is polluted.
6. Leave rubber bands out for a few more weeks.
7. Discuss their observations and inferences.

Temperature Inversion 1

PURPOSE:
The students will learn about temperature inversions and how pollutants react under this situation.

OBJECTIVE:
Students will demonstrate a temperature inversion and observe how pollutants react in this situation.

MATERIALS:
Two aquariums or two large glass jars
plastic bags
food coloring
hot plate
pin
water
ice

BACKGROUND:
A temperature inversion is a layer of cold air holding a layer of warmer, often polluted air from automobiles, industry, smokestacks, and other sources from rising off the ground and dispersing with the winds. Typically, warmer air tends to rise and cold air tends to descend, but in this state the air masses do not move; thus it is called an "inversion."

PROCEDURE:
1. To create a normal atmospheric condition, heat a pan of water on a hot plate and add a few drops of food coloring to the water. Fill one of the aquariums about three-fourths full of cold water. Add several ice cubes.
2. Fill one of the plastic bags until half full with warm water that was heated on the hot plate. Seal the plastic bag so that there is no air in the bag.
3. Remove the ice cubes from the aquarium. Lower the bag with the warm, colored water (air pollution) into the cold, colorless water of the aquarium.
4. Without disturbing the water in the aquarium, poke a hole in the bag with the pin and observe the interaction of the warm water with the cold water.
5. To simulate a temperature inversion, add several ice cubes and several drops of food coloring to a pot of water.
6. In another container, heat several quarts of water and fill the second aquarium about three-fourths full.
7. Fill the second plastic bag with the cold, colored water about half full and seal it so that there is no air in the bag.
8. Lower the bag with the cold, colored water (air pollution) into the

aquarium which has been filled with the colorless, warm water.

9. Without disturbing the water in the aquarium, poke a hole in the bag with the pin and observe the interaction of the warm and cold water.

Temperature Inversion 2

PURPOSE:
To demonstrate what happens when a temperature inversion occurs, which can trap air pollutants near the surface of the earth.

OBJECTIVES:
After having completed this demonstration lesson, students should be able to:
- Describe how a temperature inversion occurs.
- Name two primary sources of air pollutants that can become trapped in a temperature inversion.
- Understand how the activities of people interact with natural events concerning the air in our environment.
- Apply information from the demonstration model of a temperature inversion to such an occurrence in the real world, using key words: pollutant, air pollution, temperature inversion, and smog.

FOCUS:
Tell students you are going to demonstrate how weather conditions can trap air pollutants close to the ground during a temperature inversion.

BACKGROUND:
Tiny solid particles from automobile exhaust, soot from factory smokestacks, fireplaces, and trash burning are largely responsible for the formation of the haze that can be seen hovering over many large cities and industrial areas. Many people who breathe this air experience some discomfort and suffer some health problems. The severity of this form of pollution is increased when local weather conditions and/or the unique topography of a region cause the pollutants to be trapped in a layer of still air that prevents them from moving away from the area.

MATERIALS:
wide-mouthed gallon jar with cover	incense
plastic bags	plastic tubing
chilled sand bags	twist-ties
hot water	masking tape
funnel	

PROCEDURE:
1. Place a wide-mouthed gallon jar on a tabletop where all students can view it easily.
2. Place one or more very cold sandbags in the bottom of the jar.
3. Fill one or more small plastic bags with very hot water and use twist-ties to close the tops
4. Suspend the plastic bags containing hot water inside the jar by taping their closed top edges to the rim of the jar.

5. Attach one end of a length of plastic tubing to a funnel stem and place the free end in the jar.
6. Position the mouth of the funnel over a small container of burning incense.
7. Hold the jar top securely in place atop the jar and direct smoke from the incense through the funnel and tube into the jar.
8. Observe the activity in the jar.

Usually, the air that is close to the ground is warmer than that which is found at higher altitudes. This is because there is less pressure at higher altitudes and as a volume of air expands it cools. Since there is less pressure, there are fewer collisions of molecules because they have to travel farther before they encounter another molecule. (It is the collisions that give off heat, which we measure as air temperature.) However, this is only true for a column of air with uniform density. When the density of the air at the surface is dramatically different than the air above it (that is, dense, cold, dry air is at the surface and less dense, warmer, moister air is above) then we see a warmer temperature at say 5,000 feet than at the surface. This almost always happens at night and happens frequently during the day in the winter months.

When the air is especially still at times like these, the cooler air, because of its greater density, settles close to the ground, and the warmer air forms a blanket above it in a temperature inversion. Pollutants in the air, such as smoke and soot, are also trapped close to the ground. Fog, formed when moisture in the cool air condenses close to the earth's surface, becomes smog when combined with these pollutants.

EXTENSION:
Here is another demonstration to show what happens to pollutants during a temperature inversion.
1. Fill a jar half full of salt water.
2. Pour distilled water (which is less dense) on top to fill the jar. Be careful to mix as little as possible and set the jar aside for one day.
3. The next day, use a long hypodermic needle to inject food coloring into the bottom layer. The same thing happens to pollution when there is a temperature inversion.

Visible and Invisible Air Pollutants

PURPOSE:

To try and tell the difference between visible and invisible air pollution.

OBJECTIVE:

The learner will test for visible and invisible pollutants in the air. The learner will keep a journal of experimental procedures, results and conclusions.

FOCUS:

Ask each student, "What is air pollution?" Write all ideas on the board. Through discussion, arrive at one definition. Divide students into groups to brainstorm things that pollute the air.

MATERIALS:

chart paper, measuring cups
(one of the following for each group): small glass jar, large glass jar, petroleum jelly, three bean plants approximately the same size, tap water, vinegar, vinegar-water mixture in one to three ratio, pH paper or indicator.

BACKGROUND:

The atmosphere is almost completely made up of invisible gaseous substances. Most major air pollutant are also invisible, although large amounts of them concentrated in areas such as cities can be seen as smog. One usually visible air pollutant is particulate matter, especially when the surfaces of buildings and other structures have been exposed to it for long periods of time or when it is present in large amounts. Particulate matter is made up of tiny particles of solid matter and/or droplets of liquid. Natural sources include volcanic ash, pollen, and dust blown by the wind. Coal and oil burned by power plants and industries and diesel fuel burned by many vehicles are the chief sources of man-made particulate pollutants, but not all important sources are large scale. The use of wood in fireplaces and wood-burning stoves also produces significant amounts or particulate matter in localized areas, although the total amounts are much smaller than those from vehicles, power plants, and industries.

PROCEDURE:

1. Divide the ideas from the air pollution chart into two groups of pollutants: visible and invisible.
2. In groups, students will set up experiments to test both visible and invisible pollutants. Each student must keep a record of the experiments in their journal. Both experiments can be set up and run at the same time.

VISIBLE POLLUTANTS EXPERIMENT:

1. Smear petroleum jelly on the small jar.
2. Carefully place inside large jar.
3. Decide on several places around the school where students think visible pollutants will occur. Each group should have a different area to test. Make predictions about which area will have more visible pollutants and why. Record predictions in journal.
4. Place jars in test areas for several days. Have the groups check the jars daily. Record observations in journal.
5. Bring jars to class for comparison. Observe and rank the jars from the one with the most visible pollutants to the one with the least. Assign each jar a number. Discuss why certain areas have more visible pollutants than others.
6. Mark a school map showing the ranking of areas from number five. Display the map in the hall for others to see.

INVISIBLE POLLUTANTS EXPERIMENT:

1. Divide the class into three groups. Each group sets up a bean plant garden with three containers, each container having one bean plant each.
2. Students determine and compare the pH of the three solutions and predict how the plants will affected by each solution. Record pH and predictions in journal.
3. Plants will be watered every day with 1/8 to 1/4 cup of a solution: one plant with tap water, one plant with straight vinegar, and one plant with the vinegar water mixture. Procedure is recorded in journal.
4. Observe plants daily. Record in journal what happens to each plant. Sketches may be part of the observations.
5. Compare plants and discuss observations at the end of a day, week, two weeks, or until plants die.
6. Using the observations of all groups, write a class conclusion for the experiment. Record in journal.
7. Arrive at the idea that the invisible pollutants experiment was about acid rain.

Enrichment:

- Research the history of acid rain. Include information on the causes of acid rain, when we first became aware of the problem, problems have been caused by acid rain, and measures have been taken to combat acid rain. Has the situation improved?
- Make a class mural to show the acid rain cycle.
- Post a chart for the causes of visible pollutants and what can be done to prevent them. Leave the chart up so students can add to it whenever they have an idea.

The Allowance Game

Important Information about this game:
This introduction seems lengthy, but is necessary to understanding the game.

This is a learning game. By playing this game, you will learn how the government of the United States through their representative the Environmental Protection Agency (EPA) is trying to lower the amount of pollution that goes into our air using the **Market System** rather than by setting strict limits on how much pollution businesses can produce. A **Market System** has to do with buying and selling of products and services for profit. Businesses want to make money. It is why they are in business in the first place. In this game, you will see how businesses make money by buying and selling "permission slips" to pollute, and how the less pollution they produce can mean more money in their pockets.

The **Environmental Protection Agency (EPA)** tries to limit the amount of pollution that businesses emit into the air by using the **Market System**. Instead of saying, "You can only produce this much pollution," they are basically saying, "Pollute as much as you like, as long as you can pay for it." What the government does is distribute **Pollution Allowances** into the Market to be bought and sold by businesses. **Pollution Allowances** are like "permission slips" to produce air pollution. The key is that a business can only produce as many tons of pollution

as the number of **Pollution Allowances** that they own. The government keeps track of this. For example, in this game each **Pollution Allowance** permits a business to produce one ton of pollution. If your business produces four tons of pollution into the air this year, you'd better have four **Allowances**. If you don't, you are forced to buy the extra ones from the EPA, and in order to penalize you; they make you pay an outrageous price.

The greatest thing about the EPA's plan, part of the **Clean Air Act Amendments of 1990**, is that the EPA limits how many **Allowances** are distributed. This way, they have control over the total amount of pollution that gets produced. Even better, the EPA automatically lowers the amount of Allowances that it distributes into the **market** every year. There are less "permission slips" to be bought and sold. This means that every year businesses must find ways to clean up their act and produce less pollution. And this means cleaner air!

Businesses don't mean to pollute our air. In fact, many businesses try not to produce a lot of pollution. Electric companies, for example, burn fossil fuels like coal when they create electricity that we all use. This burning is what creates the pollution in our air. Businesses try to come up with ways to create less pollution. One way is by using **Wet Scrubbers**, which are inside the smokestacks and clean the air before it comes out of the smokestacks. Businesses also look for ways to use less energy. This way the company doesn't have to produce as much energy or pollution. True, they might not make as much money by selling electricity, but the money that they don't make from their customers is usually made by selling their extra Allowances to other businesses that need them. In this way, it makes financial sense for businesses to try and produce as little pollution as possible.

The reason that the EPA's plan (to make businesses buy the "permission slips" to pollute) is working so well to clean the air is that **Pollution Allowances** are expensive to buy! It costs a lot of money for businesses to buy them, therefore, businesses try even more to produce less and less pollution. This is a way for the EPA to get businesses where it really counts: in the wallet!!

Another group that has gotten involved in this plan is the **Clean Air Conservancy**, formerly known as I.N.H.A.L.E. Their goal is also to reduce air pollution. They don't have the power to make laws like the government does; instead they are an organization that also buys **Pollution Allowances** from the EPA. How-ever, they aren't a business that produces pollution. They are a non-profit organization. Instead of using the "permission slips" to pollute, they retire them. Basically, this means that they throw them away so that businesses can't buy them and use them to produce pollution. Each **Allowance** that the

Clean Air Conservancy buys means one ton less of pollution emitted into the air. Every ton retired means that $4000 worth of environmental damage is not done.

If this sounds a little confusing now, don't worry. That is why we created this game. As you play, you will see firsthand how this plan actually works and understand that the real winner is our environment.

In this game, you represent a business that produces pollution while making the product that you sell to your customers. You obviously want to make a lot of money. By trying to limit the amount of pollution that you produce, you lower the number of allowances that you need to buy. Or else, you may want to buy some extra ones. You may find that your opponents need to buy your extra allowances to cover the amount of pollution they are producing. If they are desperate enough, you can charge more than you paid for them. and make a profit.

Energy Audit
Information Sheet

This **energy audit** information sheet will help you to assess how much electricity some of the everyday objects in your lives use and how much pollution you make when you use that energy. We take energy for granted - but when we begin to see how it affects the environment we can begin to make choices that will help to change the environment for the better. After you see how much pollution your everyday activities cause, you can think of ways to stop wasting energy. After all, it is up to all of us to save the future of the environment.

We are going to look at the different parts of your school and how they contribute to your school's energy consumption. These parts include the "Box" (space, walls, windows, ceiling and floor), heating and cooling, lighting, and equipment. We will look at each part and calculate how much energy each part uses. This way we can figure out the total amount of energy your classroom uses.

After you have gathered the information necessary, you are going to fill out the **Energy Audit Work-sheet**. You will be putting numbers through a series of instructions called a mathematical **formula**. A **formula** lets you take different numbers and do various math problems with them. The results of these math problems will provide you with the answers you need to determine the electrical output of your classroom. You will use a different **formula** to determine how much pollution that electricity creates. When you finish the **Energy Audit Worksheet** you will know how much electricity your classroom generates and how much pollution this means for the environment.

STEP 1
The "Box"

One way to think about your classroom is as a box that keeps you at a comfortable temperature regardless of the weather outside. If it is a well built "Box" you will need to spend little energy to stay comfortable. If the "Box" is poorly built, or in poor repair, then the "Box" will leak a lot of air and the inside will become more like the weather outside. When this happens the only way to keep that "Box" comfortable is to use more energy to make up for the losses. The best way to think about this is to picture yourself in this room in the middle of a snowstorm. An energy efficient "Box" will keep the cold air, wind and snow from entering the building so the heater only comes on occasionally. A poor "Box" will let the cold in making the heater run a great deal more. The poor "Box" takes much more energy to keep the room at the same comfortable temperature. To determine the role of the "Box" in electricity, you need to know the size of your Box. Knowing this, you will be able to figure out how much electricity it takes to heat and cool your classroom.

Insulation

Insulation can be one of two types:

Fiberglass Insulation - This is long strips of (typically pink) fiberglass which traps air serving as a barrier between the temperature inside and outside the Box. This type is normally used in homes. This is measured in inches. The alternative is blown cellulose, loose fibers (often fiberglass) blown into the spaces between walls and in the attic. For every inch you are short of 5" you need to add .00004 to your energy needs for both heating and cooling. For example if your classroom has three inches of insulation add .00008 to you average kWh energy consumption for heating and cooling. Similarly, for ceilings you need to add .00006 for each inch less of insulation than the five inch standard.

Insulation Board - This is thin board made up of something like a foam gel that has hardened. This is a recent technology and is much thinner than the other two types of insulation. This type of insulation can only be used in new construction as it is built into the walls. If you have this type of insulation you need to make no corrections to the heating/cooling total.

Windows

The third part of the "Box" is the windows. Your windows are going to be the major contributor to air loss in your "Box." One way to measure energy efficiency is how many times an hour does the air in a room get replaced. Good windows can keep the rate down to three air replacements (turnovers) an hour while poor windows can allow the rate to go up to twenty turnovers an hour. To figure out how good your windows are you need to know the type, the size, and the number of windows.

What kind of windows do you have?

A Single Pane Window with no Storm Windows: This type of window is the one most likely to leak air and is the most common type of window for schools built before 1980.

A Single or Double Pane Window with Storm Windows: This is much more energy efficient than the single pane of glass but is very uncommon in school construction. Typically schools that were once homes but converted to schools are the ones that have these windows. Old double pane windows were an early attempt at energy efficiency but the gas between the glass tended to leak and water vapor got in making them cloudy.

A Double or Triple Pane Window with gas barriers or high "e" film: These are fairly recent, high-tech windows. These windows use inert gas between the panes of glass as insulation or use high "e" film. High "e" film is a temperature sensitive film that covers the windows to help heat stay where it should. For example, when it is cold outside, it keeps heat inside the building and in the hotter months, it helps keep the heat away from the building.

STEP 2

Heating and Cooling the "Box"

Depending on where you live and what the climate is like, different amounts of energy are required to heat or to cool buildings. Using Table 1- Average kWh Heating and Cooling, find the name of the state where you live. The number on the table represents how many kilowatts of electricity your classroom uses in an hour to heat or cool one square foot of air. kWh stands for kilowatts per hour. This is your base number. This number is adjusted depending on how you answered questions about the makeup of your Box.

STEP 3

Lighting the "Box"

Most of the electricity expended in any school is a direct result of lighting the building. This can take up as much of 75% of all the energy a school ever uses! Logically it follows that the energy efficiency of the light bulbs used can make a big difference in the total amount of energy used.

There are typically three different types of light bulbs used in classrooms.

Incandescent lights are the ones that we most often use in our homes. If they are used in your school they are most likely hidden under a light fixture or lampshade. They are shaped like the top of an hourglass and are about the size of your hand. It takes 600 pounds of coal to light one of these for an entire school year. These are the most inefficient bulbs. They have a life expectancy of **750 hours**. Up to 90% of the energy required to run one of these bulb transforms into heat - NOT light.

Florescent lights are 1-inch thick diameter tubes that range from 6 inches to 3 feet long. They are usual-

ly hidden within a long light fixture that is shaped like half of a rectangle box. It takes about 275 pounds of coal to light one of these for a year. These are more efficient than incandescent lights. They have a life expectancy of **7500 hours.**

Compact florescent lights are opaque (you can look through them but everything looks hazy). These light either have a cube base with two prongs that form a five-inch loop or they are in the shape of smaller capsules. One of these lights will result in the burning of less than 150 pounds of coal/ year. These lights are the most efficient light bulbs of all. The have a life expectancy of **10,000 hours.** They use one-fourth the energy of an incandescent and **last 13 times longer!**

STEP 4

Appliances in the "Box"

Each of the devices you plug in consumes energy. Overall, unless you have alot of these devices, they consume very little electricity compared to the lights and the temperature control devices in your classroom (radio, TV, an aquarium, film projectors) are used for short periods of time so the energy they consume is really quite small. In order to make the energy audit easier; we are going to assume 25kWh of electricity for the entire school year to count for all the miscellaneous energy consumption in your classroom. The exception to this rule is computers in the classroom, which are often on all day for your use.

In order to simplify the calculations for computer energy consumption, we assume only two different types of computers. The first of these are Energy Star computers which are specifically created to conserve energy. They have been certified as meeting US EPA standards. These devices use less energy and are "smart" - they go to sleep when not in use. The second type is non-Energy Star computers. They are "awake" and "eating energy" all the time. Non-Energy Star computers consume much more energy over the course of a day. You can tell if you have energy star equipment by looking at the computer, printer and monitor. There will be a logo on it stating "EPA certified Energy Star" and a drawn star.

STEP 5

How much pollution does all that electricity generate?

Now we know how many kilowatt-hours your classroom uses but unless we have some frame of reference to understand these numbers, they really do not mean very much to us.

The reason the values for the energy used that we found are so relevant is because they create pollution which harms the health of people and the environment. By using less energy we can improve visibility and the health of the environment. To determine the value of this pollution on the open market, you (or your teacher) need to contact the CLean Air Conservancy to determine the current price per ton of pollution. You can also access our web page at www.cleanairconservancy.org or call us at 1-800-2-BUY-AIR. The price of pollution fluctuates, so contacting us is important in finding out the most current price of pollution.

What you can do to help:

You can do a lot to help reduce pollution! The cheapest and easiest way to reduce pollution is to monitor and reduce your own energy consumption. By using less energy you emit fewer toxins into the air. Toxins are pollution agents that are created as energy is burned. If you make some of these suggested changes in your life, YOU can help clean the air.

- Use window shades and curtains to keep the heat/cool air in your house. Open your shades and blinds to let sunlight instead of always using lights.

- Watch for drafts around windows and doors. Find their sources and fix them.

- Use the smallest watt light bulb that will do the job.

- Turn off fluorescent lights when leaving a room for more than ten minutes.

- Buy and use compact fluorescent lights. Their life span is ten times longer than incandescent bulbs (those most often used at home). They are also four times more energy efficient.

- Clean light fixtures so that they are more energy efficient.

- Instead of using automatic timers, use motion detector lighting or occupancy sensors.

- Lower your thermostat by one or two degrees - you probably won't even notice the temperature change.

- Use energy efficient transportation! Ride your bike, walk, rollerblade, skateboard, and take public transportation (the subway, metro, bus) wherever and whenever you can.

- Plant a tree next to your house or building. Trees shelter buildings from cold winter winds and provide shade to your house in the summer. Trees also recycle the air that we breath. No matter how you look at trees, they are beneficial to our environments.

- Make sure you recycle! Aluminum cans, paper, newspaper, glass, plastic - all of these things can be recycled. Did you know that one recycled can saves enough energy to run a color tele-vision for three hours? If your community does not already recycle, talk to your neighbors, call the mayor's office, talk to people at City Hall and find out why your community doesn't recycle. Let your city officials know that the community cares.

- Write to your Congressional representative. Let him/her know that you are interested in environmen-tal issues, whether it be clean air, clean water, the national parks, endangered wildlife; whatever your issue, let your congressman know that he/she should be doing something to address the problem. They don't know what the people in their district (YOU) want unless you tell them.

Environmental Fundraising Ideas

- Hold a bake sale. Bake cookies and decorate them with green frosting for the environment or decorate brownies with animal faces. Connect your goodies to the environment and create a sign or flyers to tell people why you are raising money. Ask your principal if you can set up a table at lunch, before or after school, at football, soccer or basketball games. Call your city hall and see if your town has any "Home Days" activities. Maybe you can have a bake sale there.

- Get the band to play. See if you can get a local or a student band to play at low or no cost (to you) and charge admission at the door. Invite students, parents and teachers.

- Initiate a "Can You Spare Your Change?" campaign. Ask if you can set up containers at the end of the lunch line in your cafeteria for people to donate their extra change to your collections. You can set up a poster to let people know how much money you've collected.

- Sell candy bars. Contact your local candy wholesaler (marching bands do this all the time). You sell the candy bars and can wrap an information sheet around the candy bars to explain why you are raising funds.

- Put on an Environmental Science Fair. Charge admission at the door and explain that you are raising money to fight pollution. Make sure one of the projects deals with acid rain or clean air.

- Make T-shirts/Buttons/Posters. Design your own environmental T-Shirts, buttons, or posters, and sell them at school, at school sponsored events, and in your neigh-borhoods. Ask about setting up a booth at your local library on a Saturday to raise funds and sell your designs.

- Hold a car wash. Go to the school's parking lot or ask to use a local gas station to hold a car wash. Make a sign and set up an information table. Charge a low price and make sure that people know that the funds go toward preventing air pollution. Make sure that you use biodegradable and environmentally friendly soap!

- Be Creative. There are so many other options. Brainstorm with your friends and see what you can come up with.

Topic One: Pollution in the Air

1. **N**ame four sources of air pollution.

a)

b)

c)

d)

2. **N**ame four pollutants created by electricity generation.

a)

b)

c)

d)

3. **W**hat is particulate matter? **G**ive three examples of particulate matter.

a)

b)

c)

Define the following vocabulary words. Write on back if necessary.

1. carbon dioxide (CO_2)

2. nitrogen oxide (NO_x)

Topic Two: Effects of Pollution on the Environment

1. **W**hat is acid rain? **W**here does it come from? **H**ow does it form?

2. **W**hat is one of the major contributor to acid rain formation?

3. **W**hat are some of the environmental impacts of acid rain?

4. **W**hat is the difference between tropospheric (ground level ozone) and stratospheric ozone?

5. **W**hat causes tropospheric ozone?

6. **N**ame three ways you can help limit the amount of ozone on Ozone Action Days.

a) _____

b) _____

c) _____

7. **W**hat is an Ozone Action Day?

Topic Two continued

8. What is global climatic change? What causes it? How can it be prevented?

9. What is the greenhouse effect?

10. Describe what happens during temperature inversion. Why is this harmful to the environment?

Define the following vocabulary words.

1. global warming

2. smog

3. asthma

4. Ozone Action Days

Topic Three: Regulating Pollution

1. What is the Clean Air Act Amendment of 1990?

2. Who is the Clean Air Conservancy?

3. What is a pollution allowance?

4. Name two ways that you can help control air pollution.

5. What is the EPA?

Define the following vocabulary words. Write on the back if necessary.

1. Clean Air Act

2. Clean Air Act Amendments

3. Environmental Protection Agency (EPA)

4. emissions

5. Pollution Allowances/permits

Topic Four—Energy Efficiency

1. Name three ways you can limit the amount of electricity you use.

a) _____

b) _____

c) _____

2. What is the difference between incandescent, florescent and compact florescent light bulbs? Which uses the most electricity?

3. Name four types of alternative sources of energy.

a) _____

b) _____

c) _____

d) _____

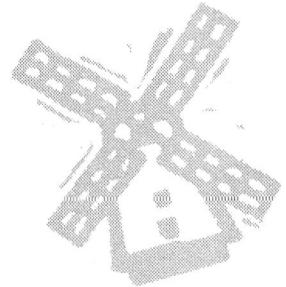

4. What is recycling?

Define the following vocabulary words. Write on the back if necessary.

1. hydropower
2. fuel cells
3. nuclear energy
4. recycling
5. natural gas
6. solar energy
7. wind power

Energy Worksheet

Your Name:_____

This is the energy audit worksheet. Follow the directions and place your answers in the correct box. If you properly follow the directions, when you are finished you should discover how much energy your classroom uses and how much pollution you are helping to cause.

Before you begin you should:
1. Make sure you read the Energy Audit Information sheet
2. Contact your maintenance director to help you answer some of the questions about the specifics of insulation, windows, etc.
3. Have a calculator.

STEP 1	
The "Box"	
1) How wide is your classroom (in feet)?	1)
2) How long is your classroom (in feet)?	2)
3) Multiply Box 1 times Box 2 to find out the square footage of your classroom	3)
Insulation 4) If you use fiberglass or blown cellulose insulation, how many inches of insulation are there in the walls? Subtract the total amount of insulation from 5" – if the number is zero or less skip to number six; otherwise write the answer into Box number four. (If you have three inches of insulation, for example, 5-3=2. Enter a 2 in #4. If you have no insulation, enter a five in number four).	4)
5) Multiply Box number four by .00004. This will be added to compensate for the lack of insulation.	5)
6) If you use fiberglass or blown cellulose insulation, how many inches of insulation are there in the ceiling? Subtract the total amount of insulation from 5" – if the number is zero or less skip to number eight; otherwise write the answer into Box number six.	6)
7) Multiply Box number six by .00006	7)

Windows	
8) Multiply the width times the height of each window to get the square footage of each window. Add together the square footage of each window to get a total square footage of windows.	8) sq. ft.
9) How much energy do you gain or lose because of your windows? The answer to this question will be used to adjust the number of kilowatts used in heating and cooling your classroom a) If you have single pane windows with no storms, then enter the number .1 here.	9a)
b) If you have single or double pane windows with storms, then enter a zero here.	9b)
c) If you have double or triple pane windows with gas barriers or high "e" film, then multiply Box number eight by .08 and enter that number here to determine your energy savings.	9c)
d) Add together 9a + 9b − 9c. This is the amount of kilowatt saved or lost by your windows.	9d)

STEP 2

Heating the "Box"	
10) Write the name of state you live in.	10)
11) Figure out how many kilowatts per hour your classroom uses for heating	
a) From Table 1 write in the appropriate number for heating (kwh/year/sq. ft).	11a)
b) Write in the answer from Box number five.	11b)
c) Write in the answer from Box number seven.	11c)
12) Add 11a + 11b + 11c.	12) kwh/sq ft
13) Multiple Box number twelve by Box three.	13) kwh
13a) Multiply Box number thirteen by 140 days (This is the number of days you need to heat your classroom. You may need to adjust the number of days to fit your school.)	13a)

Cooling the "Box" 14) Figure out how many kilowatts per hour your classroom uses for cooling.	
a) From Table 1 write in the appropriate number for cooling energy consumption (no air conditioning gets a zero for all users).	14a)
b) Write in the answer in Box number five.	14b)
c) Write in the answer in Box number seven.	14c)
15) Add 14a + 14b + 14c.	15)
16) Multiply Box number fifteen by Box number three – This is the number of kilowatts it takes to cool your classroom.	16) kwh
16a) Multiply Box number sixteen times 40. (You may need to adjust the number of days to fit your school.)	16a) kwh
S T E P 3	
Lighting the "Box" 17) How many light bulbs are in your classroom?	
a) How many incandescent light bulbs are in your classroom?	17a)
b) How many florescent light bulbs are in your classroom?	17b)
c) How many compact florescent light bulbs are in your classroom?	17c)
18) Determine the amount of KWH used by your lightbulbs a) Multiply 17a by .1.	18a)
b) Multiply 17b by .04.	18b)
c) Multiply 17c by .02.	18c)
19) Add 18a + 18b + 18c. This is the total kwh/day used to light your classroom.	19)
20) Multiply Box number nineteen by 180 days/school year.	20)
S T E P 4	
Alliances in the "Box" 21) We enter a 25 in this box to represent the miscellaneous kilowatts consumed in your classroom (aquariums etc.).	21) 25kw/h

22) How many computers are there in your classroom which are: a) Energy Star Computers	22a)
b) Non-Energy Star Computers	22b)
23) Multiply the number of computers to determine kw used: a) Box number 22a times .025.	23a)
b) Box number 22b times .15.	23b)
24) Add Box number23a plus number23b..	24)
25) Multiply Box number 24 times 180 days/year.	25)

STEP 5

How much pollution all of that electricity generates First we must determine the total number of kilowatts that the "box" uses every school year	
26) Add Box number 9d + Box number 13a + Box number 16a plus Box number 20 plus Box number 21 plus Box number25.	26)
27) Find your state on Table 2 a) Enter the state's NO_x number.	27a)
b) Enter the state's SO_2 number.	27b)
c) Enter the state's CO_2 number.	27c)
28) Let's find out how many tons of pollution your classroom generates based on the number of kilowatts used per year. a) Multiply Box number 27a times Box number 26.	28a) tons NO_x
b) Multiply Box number 27b times Box number 26.	28b) tons SO_2
c) Multiply Box number 27c times Box number 26.	28c) tons CO_2
29) How much would it cost if you wanted to purchase this pollution on the pollution market? (Check the Conservancy web page for current market prices) a) Multiply Box number 28a times current market price NO_x.	29a)$
b) Multiply Box number 28b times current market price SO_2.	29b)$
c) Multiply Box number 28c times current market price CO_2.	29c)$
30) Add Box number 28a, b, and c. This is the cost to become a Zero Discharge Classroom.	30)$

Table 1

ENERGY CONSUMPTION IN KILOWATTS PER HOUR

State	Annual Heating/sq ft.	Annual Cooling/sq ft.
Massachusetts, New Hampshire, Vermont	.018	.004
Connecticut, Rhode Island, New York, Ohio, Pennsylvania, Illinois, Michigan	.018	.007
Wisconsin, South Dakota, North Dakota, Minnesota, Wyoming, Nebraska	.2	.004
New Jersey, Maryland, Delaware, West Virginia, Kentucky, Virginia, Washington DC	.016	.01
Colorado, Nevada, Oklahoma, Kansas, Missouri, Iowa	.016	.006
North Carolina, South Carolina, Tennessee	.012	.012
Mississippi, Louisiana	.006	.018
California, Oregon, Washington	.012	.012
New Mexico, Hawaii, Florida, Alabama, Georgia, Texas	.006	.018

The numbers represent state-wide averages – your school may vary slightly.

Table 2

POLLUTION EMISSIONS IN POUNDS/KWH

State	NO_x	SO_2	CO_2
Alabama	0.00758	0.020328	2.7602
Alaska	0	0	0
Arizona	0.004	0.00678	2.034
Arkansas	0.004632	0.00874	2.852
California	0.00048	0.00006	0.75
Connecticut	0.00288	0.00288	0.00928
Colorado	0.00938	0.01062	4.19
Delaware	0.0078	0.02076	5.162
Florida	0.0079	0.01788	3.13
Georgia	0.00662	0.01922	2.894
Hawaii	0	0	0
Idaho	0	0	0
Illinois	0.00798	0.02036	2.362
Indiana	0.01382	0.0354	5.022
Iowa	0.00938	0.01816	4.248
Kansas	0.0094	0.01182	3.792
Kentucky	0.01666	0.02904	4.49
Louisiana	0.00536	0.0069	2.724
Maine	0.00062	0.00298	0.366
Maryland	0.00954	0.02286	2.928
Massachusetts	0.00448	0.0045	0.01506
Michigan	0.00756	0.0158	3.092
Minnesota	0.0084	0.00814	3.322
Mississippi	0.00636	0.00638	0.0153
Missouri	0.01086	0.02072	3.864
Montana	0.00392	0.00264	2.154
Nebraska	0.007	0.0096	2.9
Nevada	0.00756	0.00854	3.38

From *Clean Air Activities* by the Clean Air Conservancy. © 2003 by Humanics Learning

State	NO_x	SO_2	CO_2
New Hampshire	0.0044	0.00432	0.0131
New Jersey	0.0056	0.0056	0.00912
New Mexico	0.011	0.01068	4.576
New York	0.00266	0.00922	1.73
North Carolina	0.01058	0.01814	2.756
North Dakota	0.01384	0.02304	4.812
Ohio	0.01556	0.0414	3.988
Oklahoma	0.0075	0.00898	3.648
Oregon	0.00036	0.00046	0.202
Pennsylvania	0.00584	0.00582	0.02318
Rhode Island	0	0.00005	0.00002
South Carolina	0.00578	0.01058	1.668
South Dakota	0.00648	0.00558	0.3604
Tennessee	0.0116	0.0231	2.822
Texas	0.00582	0.00978	3.194
Utah	0.00906	0.004	4.438
Vermont	0.00018	0	0.202
Virginia	0.00718	0.01364	2.292
Washington	0.00076	0.00076	0.00278
West Virginia	0.01412	0.01412	0.03136
Wisconsin	0.00828	0.01618	3.78
Wyoming	0.0101	0.00948	5.062
Washington DC	0.00456	0.02714	1.714

Topic Five: Control Technology

1. **W**hat is a control technology?

2. **W**hy do industries need to use control technologies?

3. **M**ark the technologies that are used to remove particulate matter with a "P," to remove Nitrogen Oxide with an "N," and to remove Sulfur Dioxide with an "S."

_____ Wet Scrubber

_____ Baghouse

_____ Dry Scrubber

_____ Low-NOx Burner

_____ Selective Catalytic Reduction (SCR)

_____ Electrostatic Precipitator

_____ Selective Non-Catalytic Reduction (SNCR)

4. **D**oes a control technology completely clean the pollution industry creates? Explain.

5. **I**f industry is just part of the problem, what can WE do to control the amount of air pollution we cause?

Define the following vocabulary words. Write on the back if necessary.
1. Selective Catalytic Reduction (SCR)
2. Low-NO$_x$ combustion
2. Selective Non-Catalytic Reduction (SNCR)
3. Control technology
4. Overfire Air
5. Wet Scrubber
6. Electrostatic Precipitator
7. Baghouse

the clean air conservancy

Conclusion

WORD SEARCH
Search here for the vocabulary words for each topic.

```
C A R B O N D I O X I D E C R A N M A L Y A H S O O T C I M E N Y A G N J T
A L L A T R U M Z A N J E S S I I C A A N D Y O O U F G B F F G H M N B V F
S C S R Q P O N O M L K J I H G T F E D C B A L D Y F Y D H N G G D V S A S
D E T U V W X Y N Z A S D F G H R Q W E R T Y A Q W E R T Y L K H G F D E F
F N E E S U O Y E N A C Y A S H O T R E B L U R M E H T D N U H R A L L A C
G V N O W I T I S T H E S D R E G Z O N E A C E I D R A I H S O S S N F Y D
H I A B C D E F G H I J K L M R E A Q E R G T N N R G F D F H T S I T N C D
J R T G N R F V H T G F T G B K N H U H T G B E B N M I J N H O N S I S A M
K O A G H J K K Z A D B T E W T O N B M K L P R O P E L L M Y T O M M Y B I
I N D A A G E W E R A N E H W L X O O H C S T G A E R G A O T S E O G Y O K
U M E R C U R Y I S I N A L L T I H E W A T E Y R I N O H I O L J A H E N S
Y E N U K J U T H N T E D C X A D U J M W W E R D D F N M J H J E D F M E K
H N A S D R G T H Y B V D R T Y E D F Y H U J S G R A E R T V T J M U I M Y
J T A Q P L O K I M J U N H Y B G T V F R C D E Z A Q V T S D F H M G S Q H
M A Q W E R T Y U I O P A S D F G G H J K L Z X V B N M V D E T Y R H S M H
I L M N H Y T R F K W A Z D D D F G H J K L P O K I J U Y T R D E R F I B B
F P E U P E R S C R E E P E G R E E N T A X E S A S D F F U D N G E C O B A
F R E E P A C S W O T A N C V X R F M E R T Y U I O P A S D E U Y C V N N C
T O R S R L R R D A R W Y F R D D V D D F G H J K L Q W R R E R E Y D S C I
G T C E O L T T O Q E E T D E R X R E D F S G H J K L P A U J H Y C W V F D
Y E D I G T F F I W A V G T S F P V J D S E D F G H F E F W E R T L X D C R
H C F T R W E T S C R U B B E R L A N L F G T H Y U L Y H T G T R I F D S A
U T F I E H F G O E U G E T U I O X L S E V E W E C L E C V R S A N D V X I
J I G T S E R V G R B L A Y Y G K E B A W E R B U A Q W X V G H B G A B N N
I O T R S W T H E F Y T A J H H C V R A R F G N V H B C F T Y N I G Y U N F
K N Y I I A G U H G T H B T B L U G E Y T R G N A S D F V E C I T S L O S O
O A H H V Y H J T B R T R U E T J Y R A S D F M N V Z X T Y U F H J I O V D
L G J T E H J K P H E J F U M M N N F A B B E T R R D F E R G Y H J U M N G
P E U S E O K L U U A Y F T Y E A J C S F V D G D H S T A T O S P H E R E Y
V N I E R M P P P J R H S Y T V H T D F S R C P G U T E R G T H Y U T B A T
C C K I A E Q A E L G B A J R D B U T V A T A E T O R Y H B R Y G B G A S R
D Y I T E I Q W E R T Y U I L T V B Z E A W D V F G H E N J U T D B N M D E
R Q D N D S S Z K J R T H Y T R S Y S T R Y E R A U R U J K V R H Y N U Y W
F A F E I W A Q M I G Y N T G F Z H X U F U M B O E R T Y B H N J U S N V Q
G C F W X C L E A N A I R C O N S E R V A N C Y H O A E F R G Y H U J I K P
V R V C O H K C M A N T U V B V K T M Y R O N P B E A S D F U H R G H N J O
V R E G I E H F Y S T R H B Q G U M U N G Y S O R R A S V E Y J U I O P V I
C F E N D R G V T E R B T N B Y Y Y J H B O T L U T A D G H J M K I L P Q G
X V T I N E T H R D E H R M Y H H H I U P V N R O G S V C F V H B N H Y P M
A H U R O J F Y E F R B E A T B T H K O N F E E Y A B G S D F G H J K L I T
Q G L O B A L W A R M I N G R N R F L Y T T A D W R G V A E D G T J I O P R
Z I K A R F E M Y A R I R Q E J E T Y T G G R G O E T T S E F H U J I A C F
X K T R A T E J T S F V E T T A E B Y W R M T V R T M Y D R T Y J N M J U C
D V B R C Q T E R T Q F A C I T R A N E G O T O T H E S D G D G U H J B N M
```

References For Teachers

National Ambient Air Quality Standards --1996 by US EPA (1997)

Powering the Midwest -- Union of Concerned Scientists (1993)

Air Quality and Electricity Restructuring -- Center for Clean Air Policy (1997)

Emissions Trading Under the US Acid Rain Program -- MIT Center for Energy and Environmental Policy Research (1997)

Power Surge -- Worldwatch Institute (1993)

Studies of the Environmental Costs of Electricity -- Office of Technology Assessment (1994)

Renewables are Ready -- Cole and Skerrett (1995)

REFERENCES FOR STUDENTS

Amos, Janine, Pollution Austin, Tex. : Steck-Vaughn, c1993.

Armentrout, Patricia, The ozone layer. Vero Beach, Fl. : Rourke Press, 1996.

Arneson, D. J. Toxic cops New York : F. Watts, 1991.

Asimov, Isaac, What causes acid rain? Milwaukee, Wis. : Gareth Stevens Children's Books, 1992.

Asimov, Isaac, Why is the air dirty? Milwaukee : G. Stevens Children's Books, 1992.

Baines, John D. Keeping the air clean Austin, Tex. : Raintree Steck-Vaughn, 1998.

Bang, Molly. Chattanooga sludge San Diego : Harcourt Brace, 1996.

Berry, Joy Wilt. Every kid's guide to saving the earth Lake Forest, Ill. : Forest House, 1993.

Dudley, William, The environment : Distinguishing between fact and opinion San Diego, CA : Greenhaven Press, c1990.

It's your environment : things to think about, things to do from the Environmental Action Coalition ; illustrated by Susan Hulsman Bingham edited by Sherry Koehler. New York : Scribner, 1976.

Grey, Jerry. The race for electric power. Philadelphia, Westminster Press, 1972.

Herda, D. J., Environmental America. The Northeastern states Brookfield, CT : Millbrook Press, 1991.

Herda, D. J., Environmental America. The Northwestern States Brookfield, CT: Millbrook Press, 1991.

Herda, D. J., Environmental America. The South Central States Brookfield, CT: Millbrook Press, 1991.

Herda, D. J., Environmental America. The Southeastern States Brookfield, CT: Millbrook Press, 1991.

Herda, D. J., Environmental America. The southwestern states Brookfield, CT: Millbrook Press, 1991.

Kallen, Stuart A., If the sky could talk; illustrated by Kristen Copham. Edina, MN: Abdo & Daughters, 1993.

Koehler, Sherry. ed. It's your environment: things to think about, things to do from the Environmental Action Coalition ; illustrated by Susan Hulsman Bingham. New York: Scribner, 1976.

Lucas, Eileen. Acid rain education. Chicago : Children's Press, 1991.

Miles, Betty. Save the earth : an action

handbook for kids. New York : Knopf :
Distributed by Random House, 1991.

Miller, Christina G. Air alert : rescuing the
earth's atmosphere New York: Atheneum
Books for Young Readers, 1996.

Pedersen, Anne, The kids' environment
book : what's awry and why Santa Fe,
N.M. : John Muir Publications ; New York,
N.Y. : Distributed by Norton, 1991.

Shelby, Anne. What to do about pollution;
pictures by Irene Trivas. New York :
Orchard Books, 1993.

Temple, Lannis. ed. Dear world :How children around the world feel about our
environment New York : Random House,
1993.

Woodburn, Judith, The acid rain hazard.
Milwaukee : Gareth Stevens Pub.,1992.

Yount, Lisa. Our endangered planet: Air
Minneapolis, MN : Lerner Publications,
1995.

Index

Acid Rain,   3, 13, 20, 23, 24, 61
   exp. 77

Asthma,   7, 8, 50, 63, 68

Automobiles,   8, 60

Carbon Dioxide (CO_2),   3, 7, 9, 23, 24,
     56, 62, 64, 68,
   exp. 75
   chart 109-110

Clean Air Conservancy, 15,16

   History, 15,16

   Role in Sulfur Market, 15,16

Control Technologies , 17-20, 29, 35, 43, 73

   Bag Filter,   19

   Baghouse,   19, 20, 73

   Dry Flue Gas Desulfurization,   20

   Electrostatic Precipitator,   19, 73

   exp., 79

   LNB,   18, 73

   SCR,   18, 73

   SNCR,   19, 73

   Wet Flue Gas Desulfurization,   20, 73
        exp. 81

Environmental Protection Agency (EPA),   13,
     25, 33, 51, 63, 71

Fossil Fuels,   8, 10, 17, 21

   Coal,   7, 17, 18, 21

   Natural Gas,   18, 64

   Oil,   18

Global Warming,   10, 67, 70

Government Regulation   12-14, 17, 25,
    33, 71, 90

Harmful effects, 7-11, 23, 65

   Health,   7, 8, 9, 10, 11, 65

   Ecosystems,   7, 24

Nitrogen Oxide (NO_x),   3, 7, 17, 18, 23,
    24, 61, 62, 68
   chart, 109-110

Ozone,   23, 24, 65, 70
   exp., 83

   Stratospheric,   8, 9, 55

   Tropospheric   8, 9, 55

   standards chart 10

Particulate Matter   7, 10, 19, 52, 68, 88,
    exp., 83

Recycling,   64

References, 117

   Students   117

   Teachers   117

Renewable Energy   19-23,

   Project, 16-48

   Fuel Cell, 21, 72

   Hydropower, 20, 72

   Solar,   20, 64, 72

   Waste Gas,   21, 72

   Wind,   20, 64, 72

Smog,   55, 62, 70

Sulfur Dioxide (SO_2)   3, 7-8, 17, 18, 23,

53, 61, 68
   chart, 109-110

Sulfur Dioxide Market, 13, 53
   Game, 36-45

Volatile Organic Compounds, 8, 65, 66,
   68

Weather , 10, 67

  Temperature Inversion, 65,
  exp., 84, 86

Wildlife, 60

www.ingramcontent.com/pod-product-compliance
Lightning Source LLC
Chambersburg PA
CBHW082359270326
41935CB00013B/1688